Marketing Strategies
for
Design-Build
Contracting

Marketing Strategies for Design-Build Contracting

William D. Booth

SPRINGER-SCIENCE+BUSINESS MEDIA, B.V.

Art direction: Andrea Meyer, emDASH inc.
Cover design: Saeed Sayrafiezadeh, emDASH inc.

Originally published by Chapman & Hall in 1995

Softcover reprint of the hardcover 1st edition 1995

1 2 3 4 5 6 7 8 9 10 XXX 01 00 99 97 96 95

Library of Congress Cataloging-in-Publication Data

Booth, William D.
Marketing strategies for design-build contracting / by William D. Booth.
p. cm.
Includes index.
ISBN 978-1-4613-5868-8 **ISBN 978-1-4615-2087-0** (eBook)
DOI 10.1007/978-1-4615-2087-0
1. Construction industry—Customer services—Marketing. 2. Building trades—Customer services—Marketing. 3. Design services—Marketing. I. Title.
HD9715.A2B593 1995
690'.068'8—dc20 94-36928
CIP

British Library Cataloguing in Publication Data available

Contents

PREFACE

With a jolt you pull your car to the side of the road and read the job sign that can't be more than one day old: New Home of ABC Inc.; General Contractor, XYZ Construction Co.

You think out loud, "I knew the property had been sold," and your eyes roam over the real estate sign with the angled "sold" sticker proclaiming that the agent has been successful. "Thought the job would have come out on the bid list by now," you mutter as you move back onto the road and head for your office.

Later in the day you get the agent on the phone: "Jim, how did XYZ get the ABC job? I saw nothing on the builders' exchange about it."

"John, XYZ has been negotiating with ABC for the past six weeks. They did one fine job of selling the owner. Understand they just called on him out of the blue asking for a chance. They put the whole package together, plans and all."

"Didn't he get another price, Jim?"

"Sure did; two, as a matter of fact. He asked me to get two other contractors capable of doing design-build projects. He liked XYZ better. I don't know the prices, but I really feel those boys impressed him by going to him. Were you planning to bid the job?"

"Thinking about it," you answer, trying to sound very casual.

"Well, I wish I'd known you were interested. I'd like to have recommended you to the owner but I thought you mainly bid work off the exchange. I'll call you next time I have a buyer who wants to negotiate."

"Please do. Say, Jim, how about 18 holes this Saturday at the club?"

I'm sure this has happened to every general contractor at some time, but maybe it's occurring too often, and such a disturbing piece of news has prompted you to investigate the design-build-negotiate avenue of getting business for your firm.

The problem is, you are puzzled about just how to get started; after all, you aren't selling cars or TV sets but buildings, and buildings aren't sold like cars and TV sets! Or are they?

The idea of marketing construction is the innovation that progressive general contractors are just starting to realize will open a whole new field of business opportunities. A clear understanding of design-build is very important in order to pursue this new field. The design-build concept is a partnership where the contractor acts as the team leader in building the project, taking into consideration the owner's needs. The members of the team cover all areas from design and planning, to the actual construction.

The purpose of this book is to give guidelines and answers to questions on how to break into the design-build field or to improve efficiency if you are already negotiating business. Some suggestions may go against your set methods of operation and you may be reluctant to follow them, but at least consider giving them a try. Chapters 1, 2, and 3 cover selling construction in a broad and general way. The attitudes of both the contractor and the customer are discussed. In short, the first three chapters will give you a feel for what you are letting yourself in for, while the rest of the book covers the bottom-line nitty-gritty of obtaining results and making money.

You are going to approach the owner and sell your services, including saving money and time, while maybe helping the customer get the jump on his or her competition. You sell the job, not bid it. I'm not saying you won't have any competition, as other construction companies may be after the same contract. You must convince the prospect that out of all the people he or she is talking with, you are the one who can offer the best deal, and quite often this does not mean the lowest price. A personal by-product of marketing construction is the excitement and fun of putting a deal together and making it work.

Every book on selling I've ever read does nothing but state broad principles about salesmanship. The small but pertinent questions concerning the details are never answered. As a matter of fact, they aren't even asked. In this book small details are the norm, not the exception.

This book does not give financial figures. It focuses on selling the job. Working knowledge of the economics of a construction project is a must, and I assume you know your way around in that area.

There are no guarantees! When all is said and done, it will still be up to you to make things happen.

CHAPTER 1

THE CONSTRUCTION DESIGN-BUILD CONCEPT

The design-build concept will give you, the general contractor, a completely new approach to the construction market. This new approach can offer more profit per job coupled with fewer problems during the building stage, and it will create repeat work for the future. Sound too good to be true? Is there a catch? Yes. You're going to have to look at the construction business in a new light. While hard work is part of the picture, the really vital area that you will have to understand is what must be mastered mentally. Working hard is very important, but working *smart* is absolutely crucial. To that end, this chapter provides some basic education.

DEFINITION

Design-Build construction is a partnership among the construction contractors, design professionals, environmental consultants, sub-contractors and materials suppliers. Through this partnership, the design-build contractor has the capability of offering a complete construction agenda to the owner, which translates into positive benefits for both the design-build contractor and the owner.

The contractor will benefit from stronger relations with his partners, which will enable him to provide quality service in a more timely fashion. The owner will be pleased with the hassle-free, maximum value construction provided by the design-build team and will become a good source of future business for the design-build contractor.

Once the design-build team has been secured and the job won, the owner becomes part of the team. The owner can then utilize the contractor as a single sourse of responsibility for the job. The owner's business concerns will be addressed more quickly and effectively, making the entire process go much more smoothly. This will then translate into more efficient use of the owner's investment.

SALES TERRITORY

Design-build business is not something that you can do in a hit-or-miss fashion. It is not to be pursued when things are a little slow in the bid market. To be really successful, design-build should become a full-time part of your business, and then it may become the main area from which your contracts come—and in some cases, the only area. I'm speaking from personal experience. I do no formal bid work; all of my business is generated from the design-build concept.

The overall strategy or goal for the design-build contractor is to create a *sales territory* in which he or she has a large number of customers who depend on the contractor for certain goods and services. A construction contract is never viewed as a one-shot deal but rather as one of a line of ongoing contracts for the customer. Granted it can be years between building contracts, but you should treat the customer as if another contract were coming next month. You should have enough customers in your territory so that something is always coming up.

I have three good examples. I have had five contracts with a local seafood processor (three buildings and two interior jobs), three contracts with a metal-fab shop, and three contracts with an auto repair business. Note that three customers gave me eleven contracts consisting of eight buildings and three interior jobs. Most important, in each instance, after the first job had been sold, all the repeat work was contracted with me as the only contractor in the picture. Also in each case, no one (and especially my competition) knew anything of these jobs until after I had picked up the permits.

In my opinion, repeat work is the best result of design-build contracts. It's sort of like a bonus. While a second job for an owner is always nice, I really enjoy it when an owner recommends me to someone else.

You're going to have to look at marketing from the selling side, and then there will be new business attitudes to consider and follow. You'll also have to go out on the street and find the person who might need some construction, and this person has to be qualified, so you'll know exactly what and with whom you're dealing. Next you'll have to convince this person to let you build for him or her. Then comes the paperwork, and finally you can build the facility, hopefully for what you estimated, and end up with money in your pocket.

The rest of this book covers all these areas in detail and helps you get started building a territory that can be any size you want—local, statewide, regional, whatever works well for you. Of course, if you have a pre-fab metal building line, then the idea of territory may already be defined for you.

DESIGN-BUILD CAPABILITY

In order to sell an owner on the design-build concept you will have to have design-build capability. Don't think for a minute that you will be expected to put on your payroll the needed people to have design-build capability. The size of the construction company has nothing to do with design-build capability. Granted that it's convenient having the needed people in the same office, but the one-person operation with a secretary can have the same capability as the large office full of experts.

Design-build capability will require new business attitudes, in-house or outside professional services, and the creation of a base of subcontractors you can bring in to help pull together the data necessary to sell the job. In other words, you'll have to make a construction team where you're the captain as well as the coach, and as captain and coach it's your responsibility for this team to be a winner.

Everything from here on has only one purpose: to help you become a winner in the design-build section of the construction industry.

DESIGN-BUILD TEAM

At this point you have a prospect interested in the design-build concept, and it is now up to you to provide the necessary information to proceed with selling the project. You must use your design-build team to pull together the numbers.

First, someone (be it you or an estimator) has to have a very detailed understanding of what goes in a construction project. I'm making the assumption that you, the owner, will be the team captain.

We've now nailed down the most important person on the design-build team, and along with this comes the responsibility of putting all the parts together to make the sum total. This team will have architects and engineers. Your relationship with the design people is very important. The design-build contractor is selling professional services as well as construction work, so it's vital these professionals are comfortable with the team arrangement. These are the people who protect the public and will determine exactly what must be done to adhere to all building codes and regulations. This fact cannot be stressed enough to the prospect during the selling stage.

In my area certain commercial buildings that are not large do not require an architect or engineer's stamp. However, I strongly stress to my prospect that my plans will be provided and stamped by a professional no matter what the size of the building. Also, if I know my prospect is talking to a competitor, I push the point that this competitor should make it clear exactly who is providing the plans. This is a good selling tip to show your prospects that in dealing with you, they are in effect dealing with professional design people as well. So an important part of your team will be the design people, and I am sure as a contractor you have enough contacts in the field to work out some arrangements.

The other part of your team will be people directly involved in the building trade areas, that is, the subcontractors and materials suppliers. These people will provide special help for a particular area of expertise.

As the pricing process gets beyond a simple square-foot budget, which is where the numbers start for the prospect, particular information and pricing has to be provided; for example, there could be a site problem and the site work people can be brought in to help.

Thus, the design-build team is made up of a team captain, the design professionals, and the building trade people. When necessary, other experts may be included, for example, environmental or hazard waste specialists. Keep in mind your team will never be fixed, but always in some state of flux depending on the requirements of a particular project. It will be up to you as the team captain to put the team together and make it work smoothly and effectively.

ENVIRONMENTAL CONSTRUCTION

Environmental concerns are starting to impact the business community very strongly. I am sure that these concerns and requirements will become even more demanding in the future. Therefore, I think environmental construction will offer a very profitable area for the design-build contractor to pursue. The reason for this is the extremely difficult road that must be followed to obtain the necessary permits to construct anything that has an environmental problem.

What the design-build contractor can offer the business owner is single source responsibility in an area that can be completely overwhelming during the plans, permits, and building stage of the project. I firmly believe this area has limitless possibilities, but the design-build team is going to have to be expanded to include the environmental professionals to take advantage of the opportunity.

Let's go through an example. You are talking with the owner of a manufacturing firm that uses special chemicals and stores them in 55-gallon metal drums on the ground behind the building. The environmental protection people have checked out the situation and told the owner to get the drums off the ground and into an environmentally safe storage building. The owner knows he had better comply.

Now to digress for a moment, how you happen to be talking to the owner is also important. Did he call you from the yellow pages, or did he get your name from a friend; maybe you made a cold call to offer this special service. Anyway, the chances are very good you are there because of what you have done to promote business. In other words, you made it happen and that's very important.

Back to solving the problems. You determine exactly what kind of chemical compound you are dealing with, and how much has to be stored at any one time. Then you note exactly how the drums are handled; are they rolled around by hand, maybe a forklift is used or a manual hand trolley, will they be stacked or racked, and so on, until you understand exactly what will be taking place in the proposed storage building. Next you work out if the building is ground-level or truck dock height, and exactly what the dimensions are. Once you have all your data, you tell the owner to give you a few days to pull everything together.

Your next undertaking is to meet with the environment professionals that are on your design-build team and pass on all the information. Let them do a little homework and provide you with environmental information that affects construction. For example, the chemical compound is explosive; therefore, explosive-proof lighting and blow-out wall panels may have to be used, but the real problem is leakage, and all precautions must be made to contain any leakage so it cannot get into the water table. This means curbing placed in the concrete floor with a clean-out depression so leakage can be collected and cleaned up. Also, the concrete will have to be sealed.

With this information in hand, you check with your team architect and merge the construction requirements with the environmental ones. You can then work up a budget to get things going, but be forewarned about the price—it can really get up there. The example I have used is one I recently priced, and cost was three times what a normal storage building of the same size would have been.

As this example shows, the absolutely vital information is provided by the environmental professionals. Also, I might add that they may have to be brought in during construction to check or even certify a particular area. These people can become a very important part of your design-build team. They can also become the source of a lead that can turn into a nice job.

The environmental study is done before anything else is done, even before buying the building site. So the environmental people can then recommend a firm with a track record in environmental construction that will make things move faster and smoother for the owner. I strongly suggest you give this special area of construction some thought and make the environmental professional a part of your design-build team.

PROJECT STEPS

Your firm is now in the design-build business; you have convinced the prospect you are working with that you have design-build capability, and it's time to start putting together the numbers to see if this project can be signed up. There are four basic guidelines to follow.

The first step is to grasp exactly what the project is: building size, freestanding or an addition, conventional building materials or metal building, and so on.

The second step is to factor in the items that pertain to this particular project. For example, are there any foundation problems to consider? How about architectural treatment? Mundane things like overhead doors and loading docks? Then there will be office layouts with maybe a private bath for the owner. All the information you pulled out of the prospect is used here. You add it all together, make some preliminary drawings (nothing fancy is needed here, just simple line drawings will do nicely), and start plugging in the numbers. I find square-foot cost very handy. At this point you begin to get a handle on the dollars and have a good idea of how you stand with the budget if one is in the picture.

These first and second steps can run the entire gamut of construction from a simple box to a high-tech lab, but from my experience most are fairly easy to pull together. Now here's where a good database of construction costs will really pay off. You'll be able to pull together the needed information in a minimum of time.

The third step is to bring in special help for a particular area of expertise. For example, the prospect wants a special overhang to protect his customers as they use the front door. So you fax a sketch to the team architect for comments. I am not talking about a lot of time, but simply a few minutes invested in the sketch by the architect and then you get the information.

You may have to touch base with the engineers and then the site work people. Next there is electrical, heating, and air-conditioning people, and the plumber must always be brought into the picture. You may feel your database can cover all the sections for you at this point, and in a few cases you'll be correct; but from my experience, I've found every job to be a little different. To cover myself, I don't ever hesitate to bring in help.

Step three is where the design-build concept starts to get a little tricky. With steps one and two, you and your employees were the only people involved and you had complete say over what happened. Now you're going to bring in outsiders and get them to donate their time to you for a project that may not be built.

During step three the subs are important because you are

working from sketches and line drawings, not finished plans from the design part of the team. That will come later after the job is signed up. For now you are working on a budget estimate to get the selling process moving, which is where the sub comes in, and you may want to bring in a supplier to furnish information and prices. This will be especially true if a metal building is involved.

You will be mostly using the subs at this point, and you're going to have to have complete confidence in them. When the electrician tells you, based on the building use and shop equipment loads, it's going to take an 800-amp panel and it's going to cost $X, it's important to feel comfortable with this information.

Let me emphasize that right here is where the design-build concept becomes a two-way street for every member of your team. The people you want to work with will have to know that if they work with you at no cost and provide information, they will do the work if the job sells. And that, frankly, is all there is to it.

Your job as team captain is to pick the subs you have confidence in, and then sit down and work out what you all can expect from the association. It may be entirely new to you to sit down and discuss business with a sub. You may think that to deal with subs requires you to yell questions in the phone while you reach for your antacid pills with your free hand. Bid jobs seem to bring this scenario out, but on the other hand, and as I stated, design-build means fewer problems during construction, which means better relations for the general contractor with his or her subcontractors.

There is also a positive side effect for the sub, especially in the electrical, HVAC (heating, ventilation, and air conditioning), and plumbing fields, that can come from a design-build job. This is the service account. Design-build jobs are people jobs. That is, the owner and the contractor have a close working relationship and the major subs also become part of the picture.

For example, the owner will bring up an electrical change his shop foreman has suggested, so everyone meets, and in short order the electrician is on stage. If the owner likes what you have provided, the chances are very good when service work is needed this electrical firm will be called in. I've seen this happen many times. If the sub in question does not pick up on this, then point out to him that it's to his advantage to make that owner glad he's on the job. This makes three things happen. You look good by having first-class subs, the owner is happy about the quality of people he has on his job, and the sub is creating a potential service account. Everybody wins!

Added to these three positives is the fourth that evolves with the sub picking up a service account. You get a source of information on the inside who keeps you up-to-date on future work. All it takes is a phone call from your electrician telling you he was just in so and so's to look at moving some wiring for a new piece of equipment that may require more shop space. You may have it already covered because the owner was very satisfied with your work, but nothing is guaranteed and you can never have too much information about future work.

In the team concept there is another trade specialty that is very important, especially in commercial work, and that is the overhead door people. Many times they are the first contact when a business is thinking about making some changes, and often some form of construction work will have to be done to accommodate the new door arrangement. When your overhead door team member knows he has someone to call on to handle the necessary construction work, this means good business for all concerned, plus you the contractor will have another company contact that may produce a future construction project.

You set out to create a business association of trades that you can trust and depend on, as well as feeling comfortable with the information they provide. I stress the team concept. Often after I have met with the electrical and HVAC subs, they work together to define the areas where they interface; and it's done in their office, not mine. It will be an advantage if you can take personality into account when looking over your list of subs because this association will have to be made up of people who can work together.

To quickly reiterate, step one is to pull together a general overview of the proposed project. Step two is to factor into your overview all the special requirements for the job. Now frankly, steps one and two go hand-in-hand, and in many cases will sort of merge together. Steps one and two will generate the general specifications (loosely used here) and a preliminary line drawing that in turn will be used to bring the needed experts into the picture. Step three is to bring in team members to provide budget estimates. When you have all the outside figures, you can then proceed to the final step.

The fourth step is to establish a price and present your proposal to the prospect. This proposal, be it a beautifully bound pamphlet or a page from a yellow legal pad, is the starting point for the project. To finally end up with what the prospect wants may require redoing the specs and drawing. Make sure everyone understands that the first pass at the job will probably not be the only one.

Don't be discouraged at this point. I know it reads as if days and days will be necessary for you to work up this budget estimate. But once you have a couple under your belt, you'll be pleased to discover the time involved is hours, not days. With all the marvelous gadgets we have today you can carry on your business from anyplace. For example, you can fax the specs and line drawings to your subs and they can return their figures the same way. Or, if you're sitting on the job site with some time on your hands, pick up your car phone, open your briefcase, and go to work.

One bit of advice. Start with a small job; don't jump into a 30,000-square-foot building. Begin with 5,000 square feet and work your way up as you become comfortable with design-build.

Remember you're going to have to make something happen. The design-build team is only as good as its captain. Somebody has to be responsible for the end result, and that somebody is you.

Let me make a point here concerning the design-build concept. Everything that has been covered in this chapter is construction oriented, and general contractors should feel comfortable with this because it is their field of endeavor. While some ideas or suggestions may take some thinking before they are accepted and tried, everything is still in the construction field. However, from here on the general contractor will be stepping out of the construction industry into marketing. That is what the rest of this book is about: *taking the design-build concept and selling it.*

Many times in the coming chapters we will refer back to a subject in Chapter 1 and show how it fits in with the selling process.

CHAPTER 2

THE ART OF SELLING DESIGN-BUILD CONSTRUCTION

Selling is an art, not a pure science. It is dealing with people; thus, there are no ironclad rules. The rules change with the individual, so the person practicing the art of selling has to be able to adapt to the situation.

Selling is simply one person setting out to convince someone else to buy his or her wares. Everyone who is in the process of earning a living sells. It can be the man behind the counter of the local hardware store selling tangible goods, or it can be the insurance agent selling intangibles. The secretary is selling skills. The doctor thumping your chest is selling knowledge. The lawyer sitting quietly listening to the details of a potential lawsuit is selling services. When the general contractor puts in a bid, the idea is to sell ability to perform the work, at the lowest price. This is selling as thought of in its broadest sense. Many people do not think of themselves as salespeople pushing their goods and services because this idea may not be an accepted part of their way of earning a living. The term that I like to use for the selling of those people who sell but who don't recognize the fact that they do is *passive selling*. Office workers, military personnel, and factory employees are all passive salespeople, making up by far the bulk of those in the job market.

The other side of the coin is represented by the person who accepts the fact that selling is one of the most important business functions that he or she can perform—the aggressive salesperson. And I don't mean the "foot in the door" type of aggressiveness, but rather the "go to the customer and ask for the business" type. Many people practice the art of aggressive selling. You see them every day: the building supply salesperson who calls on customers regularly, car dealers, shopping center merchants, and general contractors—they're all out in the marketplace practicing aggressive selling. They are trying to get the customer to spend his money with them. The result of this effort is the greatest success story the world has ever known: free enterprise.

The goal is to help the general contractor bring his aggressive salesmanship to its highest peak by going out in the marketplace and successfully selling his skills, knowledge, and abilities to an individual in need of those services. Now this is an entirely different type of selling from what the general contractor is used to. Most of his background and experience have been in bidding for business, but now we are talking about going and asking for the chance to build a new building. The person who does this, be it the owner or one of his employees, will have to become a true, honest-to-goodness salesperson in every sense of the world.

ON AND OFF BUTTONS

The first point a salesperson must understand is that people have *on* and *off* buttons, and knowing how to push the *on* button is the first step in ending up with a signed contract. At the start you are not worrying as much about the *on* button as you are about taking pains not to push the *off* button. For example, I'm sure you have at some time walked into a clothing store, let's say, looking for a new sport coat. You quickly find the sport coat section and about that time the clerk comes up and asks to help. You tell the clerk that you'd like to browse for a bit; the clerk proceeds to ignore your request and hovers at your elbow telling you how this coat or that coat is made for you. After a few minutes of the pushy selling approach, you leave—not because there was a poor selection of sport coats but because the clerk pushed your *off* button.

Think back over some of the times a salesperson has turned you off because of the manner of his approach. Now think about the good experiences, when you have said to yourself, "I like that person and he knows his business." Chances are, the differences are in the sales personalities of the individuals with whom you are dealing. The way a person comes across to you at the start creates an impression that is almost always permanent.

When you call on a prospect, no matter whether it's a cold call or one that was recommended, you want to create a good impression. You want to come across as a nice guy, thoroughly grounded in your business, who thought enough of Mr. X's business to come in and ask for it.

Whenever possible, you should call for an appointment. This usually happens when you find that Mr. X is in the market for a new building and you want to contact him. This phone call does two things for you: it shows Mr. X you run a business-like firm, which is helpful in having Mr. X start to form the correct impression of you; and the call also lets you find out if Mr. X does indeed plan to build, which helps you make better use of your time.

I recommend that when calling a perfect stranger on the phone for the first time you state your name, your company name, and then ask if that person has a few moments to talk to you. One of the rudest assumptions is that the other party has time to talk just because he's on the other end of the line. So ask if it's convenient, and if not, say you'll be glad to call back. And ninety-nine times out of a hundred, Mr. X will tell you to go right ahead. So you've cracked the door, and Mr. X hopefully will start to form that favorable impression of you, since, by asking for his time, you have told him, "I know you're a busy man and we're having this conversation because you want to." You have put Mr. X in charge, and everybody likes to be there. After the preliminaries, get to the reason for the call and then say goodbye.

Many times it's not practical to phone ahead, for example, when calling on a likely prospect whom you know nothing about. But it makes no difference whether you have an appointment or are making a cold call. The idea is to not to push the *off* button.

When you have an appointment, be on time! If for some reason you cannot make it, call as soon as you know and do give a reason. Mr. X is in business and has problems, so he'll understand when you tell him a truck backed into a freshly laid block wall and you have to be there.

Now remember the only goal you have set so far with Mr. X is not to push his *off* button. You are in essence simply selling yourself, and you are the first thing Mr. X has to buy, long before he signs the contract.

You arrive at Mr. X's place of business, report to his secretary that you have an appointment, and finally meet Mr. X face-to-face. But hold it a minute—suppose Mr. X runs an auto parts store and there are five customers waiting. Then you stand to one side out of the way until Mr. X has everything under control and wait until he can see you. Then he may leave you to wait on a customer. Mr. X will always put you second to the customer if he is operating a walk-in type of retail business—a fact you'll have to live with. The point is, make yourself fit into Mr. X's scheme. Sometimes it's not easy, but it is necessary.

Regardless of what you have done to meet Mr. X, you are finally there; and, as is customary, the two of you shake hands. Now ask yourself: What did Mr. X think of my handshake? This is important because I firmly believe more *off* buttons have been pushed by "limp fish" handshakes than all the other reasons combined. If you don't know what kind of handshake you give, then I seriously suggest you find out. Nothing makes a better impression than a brief, firm handshake.

Now that the meeting is under way, you the salesperson, will have to do the initial talking to get the interview off the ground. After all, you've come to Mr. X, and he wants to know the details of why you are there. As the meeting progresses, the good salesperson starts to control the conversation. Note that I said control, not dominate. People seem to have the preconceived notion that the hotshot salesperson is a smooth, fast talker who never gives the buyer a chance to say a word. Not so!

The really successful salespeople are the best *listeners*. The first item of importance in the interview is to learn. You need to find out what the prospect's plans are: Does he have a site, has he obtained financing, and can you help in any way? Are there any special problems that must be looked into? Learn all you can; you can never have too much information about a prospect. All this evolves by asking the correct questions—and then sitting back and listening. The impression made is one of trying to be helpful, not one of telling the prospect that you have just the thing for him. Rather, you're asking what the requirements are and how you can help by using those requirements.

By the way, never, never, tell Mr. X how to run his business; that's a sure way to push his *off* button.

There are three particular *off* buttons that I feel strongly about: the beeper, the car phone, and the office phone.

Nothing is as distracting as a beeper going off right in the middle of a meeting with a prospect. And I'm not the only one who feels that way. Ask around, and you'll discover (as I did) most people don't like it. So if you use a beeper, take if off and leave it in your car. Even if it's turned off but still on your belt, the prospect doesn't know this and he or she will be waiting for it to beep.

If you have a car phone and your prospect is in the car to see some examples of your work, turn it off. You have a captive audience, so take advantage of it. Don't think you'll impress the prospect with how important you are, as you tool around town, because the office simply has to get hold of you. Forget it—he or she will not be impressed. Instead the prospect will think about what's happening back at the store, and then find some reason to cut short the trip.

If the prospect comes to your office for a meeting, hold your calls. If no one is up front, turn on the answering machine.

It's hard enough to get and keep the prospect's interest, so when you get the opportunity, do everything you can to improve your chances.

LOW PROFILE AND SOFT SELL

At this point what the salesperson is doing is maintaining a low-profile, soft-sell attitude that does not put off the prospect. The salesperson is selling himself and making Mr. X start to think about doing business with him.

Low profile, soft sell is the exact opposite of the obnoxious, loud-mouth, contract-waving, slick talker. Low profile and soft sell means doing your selling in a courteous, respectful, businesslike manner. You never presume to tell the prospect what he needs until you have all the facts and he asks you for your opinion. You are helpful, and you don't get bent out of shape when you are inconvenienced by the prospect because of the pressures and responsibilities of running his business. You understand the prospect's particular problems and are able to fit yourself into his picture. You sell to the prospect in the same way you would want to be sold to if the roles were reversed.

Many useful things can be found out during the first meetings with the prospect, but most of them are technical and will be covered in detail in later chapters. Right now we are interested in the general, overall selling picture and how it works for the construction salesperson.

The construction salesperson has one important thing working for him that puts him right at the top of this profession: he is selling goods that are not purchased unless they are needed. He never pushes something off on the customer. When he sells a building, it's needed. This gives the construction salesperson a mental attitude about his work that allows him to think of himself as truly a professional in his business community. The reason is simple. Mr. X might buy some $10 gimmick for his car when he's not sure he needs it; but he'll never buy that 5,000-square-foot storage building unless he really needs it, no matter how good the salesperson is. The construction salesperson provides the same services as any other professional; he fills the needs of the buyer when the buyer is in the market for his special services.

So far our selling attitude has been centered around the idea of low profile, soft sell, but the trick is to achieve this and remain aggressive. The salesperson must know just how far to carry a subject, when not to pursue a certain avenue; and this knowledge can only come from experience. Every good salesperson has to work out his own system.

The main way to remain aggressive while practicing low profile, soft sell is to show interest. There are many ways to show interest during the selling period of contract negotiations, and those ways will be discussed in detail in later chapters.

TIME INTERVAL

There is an amazing fact concerning the businessperson that may or may not be known to you: Mr. X will finally decide to build that new facility only after he is forced to face the hard fact that he must have a new building. He will hang on to his old place for dear life until the very last minute.

For a businessperson the prime mover is always money; in every decision it is the predominant factor. Why should Mr. X build the new warehouse and double his monthly mortgage? He will build when forced out by redevelopment, or if a new highway is going through his loading dock; business expansion is always a long and agonizing undertaking. The reason I mention this now is to point out that during the selling process it may take months or even years, in extreme cases, for Mr. X to get to the point where he is ready to sign up and get the ball moving. Construction salespeople must accept this fact. Very few sales are made quickly; most take between two and six months. And if you are doing your job effectively and efficiently, this is the ideal time sequence. The idea is to make contact with the prospect in the initial planning state and carry through to contract signing, then on to the building stage. Since most businesspeople move cautiously, it will take time to do the selling job right.

A quick project, that is, one where the prospect has done a great deal of preliminary work and wants to get started right away, is a project that is already out on the street. The end result is one that boils down to the low price, and that is what you are trying to get away from by selling construction projects.

Obviously, then, for the first months you or your salesperson is operating, there will probably be no contracts coming in. It will take time to get prospects in the pipeline so that there are several being worked on at various stages; then before long the repeating process will take over, where jobs are dropping out as new prospects are added. Just bear in mind that if you end up with some fast business, it will be more luck than anything. So don't be discouraged when you start selling and get the idea that you are doing nothing but spinning your wheels.

EMPATHY

A good construction salesperson should be able to practice empathy. Empathy is the ability to put yourself in someone else's place and see things from his or her point of view. From the salesperson's side it may look like Mr. X has no choice but to build, so you get in gear and go after the job. Then Mr. X tells you why he's not going to build; and unless you can change places with him, it may be hard to understand.

Empathy can also make your job easier. For example, you get the feeling a prospect is putting you off; he's not doing enough to push the project along even though both of you know he plans to build. Looking at the situation from the prospect's viewpoint may show you he really does not have a time problem and is not pushing you around.

So far we have talked about the fundamentals with which the construction salesperson should be comfortable: practicing low profile, soft selling, and yet remaining aggressive; thinking about how to push *on* buttons and how not to push *off* buttons;

being available at the convenience of the prospect; learning to listen; impressing the prospect with your interest in his business; accepting the sometimes excessively long time intervals; practicing empathy.

All of these suggestions are only guidelines, not ironbound rules. As I said earlier, selling is an art, not a science. Every individual who undertakes to sell construction projects aggressively is going to have to work out what is best for him or her. All I am saying is here are some important areas that should be given careful consideration when you are eyeball-to-eyeball with Mr. X.

As you read on, I will refer you back to these basic guidelines with specific details to help you get a better grasp of them.

UNDERSTANDING THE BUSINESS PROSPECT

At this point I would like to change the subject and consider the prospects whom the construction salesperson will be calling on by business category. There are three main areas: retail, wholesale/industrial, and professional.

Retail

The first thing that you will notice about the retail segment of the business community is they think small (I don't mean this in any derogatory manner at all). The retail merchant for the most part sells items that cost from pennies up to several hundred dollars; he's used to dealing with small sums of money. At the same time his business outlook is shaped on a day-to-day basis.

Prove it to yourself. The next time you visit your local hardware store or drugstore, ask the owner or manager how business is. His answer will probably run along these lines, "Not too bad today, but yesterday was really off." To the construction person who thinks in terms of good and bad years, this day-to-day existence is a little difficult to fathom, and sometimes hard to deal with.

When the retail merchant has a bad day, he has a hard time spending money; and believe me, one bad day can ruin the whole deal. For example, I was working with a retail merchant who had been leasing for years, and finally reached the point where he was ready to take the plunge and build his own place. Business was good, and after three months of preliminary work he told me to draw up the contract. The very next afternoon I stopped by for his signature and was startled when he told me he wanted to hold off for a while. The man sold toys, and that particular day was awful. Based on that day he decided to postpone the project until after he saw how his Christmas business would be. Well, I ask you, if a toy merchant can't look forward to and expect good business during the Christmas season, just when does he expect it? The man never built; there was always some reason not to.

Timing is very important when you are calling on and working with the retail businessperson, so make it a point to find out the best time to make contact. He will appreciate knowing you don't plan to take up his busy times.

It is maddening at times to try to carry on a meaningful conversation with the retail merchant when he's watching his sales area; and even if he uses his office or walks back to the stockroom, he is still subject to interruptions. Remember, as much as he may want to talk to you, his customers will always come first.

A method I have found to be most useful is to get Mr. X away from his place. Lunch is a fairly productive solution, but there are always distractions buzzing around. The best way to get Mr. X away from his clerks and the phone is to take him to look at a building that is similar to what he has in mind. The prospect then becomes a captive audience and the salesperson can accomplish a great deal in a short time. Another good method is to arrange a meeting after hours, say in the evening or over the weekend. The important point is that the retail merchant is at times difficult to talk with, and the good construction salesperson is going to have to be able to maneuver the prospect into a meeting that will pay dividends.

Let's go back to the money angle. As already stated, most of the time the retail merchant deals in small sums of money, especially when compared to contractors. Say $200,000 to the average retail merchant and notice the color slowly drain from his face and a slight glaze cover the eyes, whereas the general contractor is used to running large sums of money through his books. The average contractor thinks no more of $200,000 than he does of $2 million. The trick in the construction industry is to have some left after running these large amounts through the books. Right?

The retail merchant who is in the market for a new facility will generally have an idea of what the new building should cost. The figures will probably be at least 10 years old. So when you give him some budget numbers, Mr. X will gasp and say, "I had no idea of spending that much; it's out of the question. I'll just stay put for a while."

At this point the construction salesperson has to educate the prospect and slowly get him used to the numbers. This takes time and persistence. Remember, this money to the contractor is really not very much, but to the average retail merchant it's probably the largest money transaction he will ever undertake. He is reluctant and apprehensive. The salesperson must understand this, accept the fact, and work toward making the prospect comfortable with the figures.

Sometimes the retail merchant does not have the financial background concerning large sums of money that you might expect him to have. He has no idea at all about the details of pulling together a financial package to pay for the new facility. Here the construction salesperson can be most helpful and, at the same time, make himself more important to Mr. X—which is the name of the game at the selling stage. To sum up, the construction salesperson must keep in mind when working with the retail businessperson that this prospect is the exact opposite of the construction person in the manner that business is viewed and conducted.

Wholesale and Industrial

Wholesale and industrial is as different from retail as day is from night. This is the segment of the business community where you as a construction person will be able to establish a very good rapport.

Like the contractor, the wholesale and industrial businessperson is used to working with larger sums of money. This person does not freeze up when the salesperson starts throwing around $200,000. Our prospect is right at home. The wholesale and industrial businessperson does operate his firm on a daily basis as we all do, but he thinks in terms of good and bad months, quarters, and years. Also, within this time frame he can have seasonal peaks and valleys, like the contractor. Having to operate in this manner requires planning; therefore, our prospect will be more willing to listen to a good sales presentation and to appreciate ideas and help on future expansion.

Our prospect will have more employees than the retail person, including, in a great many cases, delivery personnel. He will have all the problems that go with having people on a payroll and trying to get the goods delivered to the customer.

The contractor faces the same situations trying to get his job done, and will have no trouble establishing a line of communication with this type of prospect, and through this, acceptance. Several years ago I was working with a good prospect who was in the wholesale food distribution business. I had already made two calls on this particular gentleman but felt that I was not really getting through to him. He seemed distant and noncommittal. I was having trouble obtaining the necessary information to proceed with working up a presentation. I hung in there because the man was a first-class prospect.

The bank located next to his location had bought the prospect's warehouse after many months of negotiating. He had been given six months to move. Two months had already passed when I got into the picture, so Mr. X was very quickly running out of time, a factor that made him a prime prospect. During the second meeting I had determined that activity eased up during the middle of the morning, so I had a third meeting scheduled for the following Monday at 10 A.M. When I checked into my office, it was a typical Monday morning, and one of our truck drivers had stayed out, creating a bad problem on a special job. I had my secretary call, explain my problem, and push back my appointment. I finally solved the problem and showed up 45 minutes late. The first thing Mr. X asked when I stepped into his office was if I had much trouble with drivers staying out on Monday mornings. It had been a hectic morning and I must have been looking for a sympathetic ear to bend. I took 10 minutes and told him how it was really difficult to plan for Monday mornings when you didn't know who was going to show up.

After my tale of woe, Mr. X sent for coffee and proceeded to tell me about his morning. Two of his drivers failed to show. We spent the next half hour telling each other how hard it was to find good help, then the next hour working out the details of his new warehouse. He was a completely different person. Suddenly I wasn't an outsider pushing my way in, but a kindred spirit—one that could be talked to and counted on to understand. Two weeks later I had a signed contract for a 10,000-square-foot warehouse, with a 2,500-square-foot attached office.

The mechanics of dealing with the wholesale and industrial businessperson are much easier than dealing with the retailer. The former operates from an office behind a secretary. You will be able to make an appointment and conduct your business over a desk in an efficient manner. This helps the salesperson because it means he will be able to put his time to better use than standing around waiting for the prospect's customer to clear out. It is easier for the salesperson to drop by and hope to catch Mr. X in; and if he finds Mr. X in, the chances are that Mr. X will see him. Since he is not directly involved with the customer, our prospect has more control over his time.

After making several contacts, the construction salesperson will come to understand and appreciate the wholesale and industrial prospect from still another point of view: *the size of the job.* These businesspeople need space to carry on their enterprise: the wholesaler needs space to stock his goods, and the industrial person has to have space to manufacture his wares. These people think in terms of 10,000 or 100,000 square feet, and that's what makes the construction salesperson sit up and take notice.

The important thing to keep in mind when establishing a contact that you hope will result in a contract is that the wholesale and industrial prospect is in many ways similar to the general contractor, and these similarities can be used. Better yet, they should be pursued and cultivated. The sooner a rapport is built, the easier the selling.

You may be thinking, why all the fuss over finding a common ground with the prospect? What's the big deal? Just talk football or golf, and do the *same job.* Try it and you'll wonder what happened. The retailer has very little to nothing in common with a general contractor in a business way, so you maintain a pleasant, professional relationship and go from there. The wholesale and industrial people do have business areas that are very much like the general contractor's and the good construction salesperson will capitalize on them. But never try to make something up with sports or whatever. Realize that these people you are wanting to do work for aren't dummies. They are astute members of the business community and didn't get to where they are without some gray matter between the ears. They'll read right through a phony "hail fellow, well met" greeting, their *off* button will start glowing a bright red, and you're out of the picture. Even worse, they might tell their friends about you.

Now there are exceptions. Remember, selling is an art, so there will always be exceptions. If the prospect brings up the subject, okay, pursue it—but not into the ground; don't keep flogging it. It's little points like this—knowing how far to carry a subject, for example—that make the difference between a good salesperson and a super one.

Professional

The professional category includes doctors, lawyers, CPAs, dentists, and so on. The general contractor has even less in common in a business way with this group than he does with the retailer. These members of the business community don't have hungover drivers on Monday morning.

They seem to rely a great deal on other professionals when they need something done. Thus, when you are dealing with the professional you cannot overstress the fact that design service professionals will be overseeing and designing the project. Don't hesitate to give out the names of your affiliates because your prospect may know them, in a business manner or socially.

These professionals are hard to see! It's almost impossible to get past the secretary on the telephone, and don't even waste your time stopping by if you don't have an appointment.

When selling you have to play the percentages, and with respect to percentages the professional category doesn't offer as many business prospects as do the first two groups.

There are exceptions: professional people like doctors, dentists, and lawyers are constantly looking for places to invest money, whatever the reason. Commercial real estate can be made attractive to them. The construction salesperson can use this fact to put together a venture with the clear understanding that he is going to build the facility. If this segment of finance is new to you, then I suggest you make it a point to learn all you can because it will be a big help to selling.

To summarize briefly what we have said so far, the methods that the construction salesperson can use to gain entry to the prospect are very important. It takes two to sell something, and you are only half, so you must be able to confront and hold on to the other half. We have discussed what to do and what not to do to avoid turning off the prospect. The three general areas of the business community have been looked into, and their description will prove helpful in letting the salesperson prepare mentally for the particular type of prospect he or she will be calling on. Several more topics are discussed below. They are just as important to selling as anything that has been presented so far.

Don't Talk Down To or Put Down The Prospect

This admonition will probably be obvious to the reader, but it still needs to be brought out and should always be in the back of the salesperson's mind. The construction salesperson calls on all kinds of people, and he has to be as adaptable as a foot soldier in the field. From the plush well-appointed office of a trucking company president who wants to spend $400,000 to enlarge his loading docks, the salesperson may call on the owner of a small auto repair garage. The office consists of one desk in the corner covered with paper of all kinds. It will most likely look like someone emptied the trash can on top. All the paper will be smudged with grease, and right in the middle will be two or three wiping rags. The couple of chairs around the desk can be classified as early grease pit. You hope above all that you can do your business standing up!

Most garages these days will have a guard dog, and there he is under the desk giving you a beady-eyed look that says, "You're here only because I let you stay."

Mr. X comes up and you introduce yourself and put out your hand. That's right, put the hand out; assume his is squeaky clean. If he has a dirty hand, he'll most likely say so and let you know he'd like to shake your hand except for the dirt. But if he proceeds to shake hands, then grab it. When he points to a chair, sit in it; but as far as the dog is concerned, you are on your own—the best advice I can give is to ignore it.

Granted that this description may be a bit heavy-handed, but I feel it makes an important point. If you look around as if you can hardly wait to get away, or don't sit, or maybe pull out your handkerchief to wipe your hands—then bang goes the *off* button.

My experience has been that the self-made man is a little more sensitive to the put-down, no matter how slight. I'm sure a psychologist can go on for days about why, but I really don't have the space for that. It must suffice for now that you be aware of this situation. This selling area has to be played by ear.

Don't Overlook Selling To the Prospect's Employee

Whenever you get the chance, talk with the prospect's employees. It's surprising what you can learn that may be a big help with your negotiating. For example, I was trying to sell a motorcycle dealer a new building recently. He planned to relocate and had a new site, so he was a prime prospect. I kept going back time after time to try to get the project going so that I could put together a proposal. Well, after two weeks of feeling pushed around, I was ready to drop Mr. X and move on to something else. I decided to make one last call. I phoned, set up the appointment, and showed up on time to find Mr. X at lunch. Business was slow, so I struck up a conversation with the sales manager. We talked about a number of topics; then I brought the conversation around to the new building. The sales manager was very excited about the idea of a new building. He had a head full of ways to increase the business. I asked for his thoughts on several showrooms and shop layouts that I was working on. The man was pleased as punch to help. Now, I honestly wanted his help, but at the same time I was selling myself to him. I wanted to find out if anything was happening that I should know about. I mentioned that Mr. X seemed unconcerned about moving along with the new building and was a hard man to do business with. Then the manager dropped the bomb. The building they were operating from had been condemned the previous month and they had to be out in six months. Mr. X's attitude was just his way of trying to get the bottom price.

Needless to say, I stuck it out, and negotiations are still going on at this writing. Maybe I'll get the job yet. My point is: I was ready to give up until I fished about and made a big catch, which was exactly what I intended when I started talking to the sales manager. So chat with the hired hands every so often and you may get some useful surprises.

Be Conscious of Prospect Embarrassment

This area of discussion may be a little vague and may even cause some outright laughter. But there is such a thing as prospect embarrassment. I know because I lost a nice contract because of it.

When the salesperson is working with a prospect from the beginning, there usually are many changes necessitating a constant flow of information from the prospect to the salesperson. After asking the salesperson to change this and that for what seems like the tenth time, the prospect may begin to feel a little guilty about using the salesperson's time. If the prospect is not completely sure about going ahead, then he may just back off and lose his momentum rather than ask the salesperson for something else. The salesperson at this point may not be pursuing the prospect as actively as before for some reason or another. The project just dries up for lack of communication; the prospect is too embarrassed to ask and the salesperson does not follow up.

The next thing the salesperson hears about is Mr. X building a new place, and he wonders what happened. He has no one to blame but himself, and chances are that he won't be aware of what really went on. All the fault lies with the salesperson for not making clear to the prospect that he is to call on him no matter how many changes are necessary. Make the prospect understand that it goes with your job, you expect to do it, and he should not worry about your time. The good construction salesperson should be able to tell if the prospect is serious or is just manipulating him, looking for information that can be used later. This is a judgment call, and with a little time and experience to his credit, the construction salesperson will learn to make the right call.

The salesperson must *keep* impressing on the prospect the importance of always keeping the lines of communication open, no matter how much time is involved. The good salesperson will not depend on Mr. X to do all the work. The other mistake the salesperson must not make is failure to follow up. The prospect will not mind using your time so much if you offer it face-to-face. Remember prospect follow-up is one of the most important aspects of closing a contract.

That job I lost was a warehouse for a heavy equipment dealer. It seemed everything was always being changed. I started on two other projects when I was working with the heavy equipment dealer, and I let things slip. Not hearing any more from the dealer, I thought he was still trying to decide what he was going to do. Two months later I ran into the dealer at lunch and asked him how things were coming along on his building. He looked slightly uncomfortable and told me he had bought from another firm. Before I could say anything he went on, "I felt so bad about grinding up so much of your time. I didn't want to bother you."

I stood there dumbfounded! How could he bother me if there was a contract involved? Well, I chalked it up to having to deal with individuals and the fact that everyone is different. Then six months later I was having coffee with a friend who builds custom houses and performs a great many of the same services as a design-build contractor. We were comparing notes on selling construction and he repeated a story that sounded very much like the one I have just described. He had experienced prospect embarrassment also. I have since talked with several other good construction salespeople and they all agree that there is such a thing.

If you are skeptical, that's all right. But when working with a prospect, tell him not to hesitate to call you. And follow up.

Stand Straight and Look the Prospect In the Eye

Do not feel that you are imposing on the prospect. Do not apologize for being there. Don't slip in the door and stand in the corner twisting your hat, hoping finally some nice person will speak to you in friendly tones. Walk in as if you know exactly what you are doing. *Look, be,* and *act* confident. No businessperson respects a creampuff. And if you're wondering if you can do this and maintain the low-profile, soft-sell attitude, the answer is yes. It's not that easy to explain; it's something that just has to be done. I think the best advice is to practice the low profile, soft sell, but give the impression that you are a businessperson conducting business during business hours and have no reason to be apologetic.

I like to take the direct approach sometimes when I know something about the prospect. For example, with the Mr. X who has the reputation of chewing people up who call on him, after doing what has to be done to get to Mr. X and while shaking hands, I look him in the eye and say: "Mr. X, I'm here because I can save you quite a lot of money on your new building. It'll take me five minutes to tell you how." I have never had this direct approach fail. Mr. X always takes the time to listen. What that direct approach does is reach out and grab him by the shirt. You have made Mr. X want to talk to you, and that's what selling is all about.

This direct approach will not be the best way all the time; usually it is best for the fast-moving fire-eater type. The trick is to figure out what type of person you are calling on, so find out all you can about a prospect before you meet him. Calling a mutual friend is an excellent way. If possible, observe while waiting; doing so should give you a good handle. Another non-apologetic approach is one I use when the prospect gets into money and feels maybe the price is too high. I explain the cost of construction, and then, making it a point to look him in the eye, tell him profit is not a dirty word and that I fully intend to make a fair profit out of the job because that's the reason I work—the same reason that he, Mr. X, needs and wants the new building: to make money!

I generally don't use this approach unless I get the feeling that the prospect is looking down his nose at me; then I quickly yank him back to why we are talking and the fact that we are both after the same end—making a buck! Never apologize for wanting to make a profit. It's really the main reason you are out there in the big bad world.

Remember, you are not only selling. You are operating in one of the truly specialized areas of selling. The construction salesperson not only has to be a good salesperson but has to have a good working knowledge of the construction industry. He has to be able to understand zoning, sewer and water problems, drainage, the various trades and how they dovetail together, architectural aesthetics, site work, paving, construction finances, permanent financing, and so on. In short, he has to be thoroughly knowledgeable about a very large and complex industry. At the same time, he is selling in the big league. Most salespeople close deals for $100 or $5,000, but the construction salesperson closes contracts for a quarter of a million dollars with the same ease that a tire salesperson sells $80 worth of tires. So hold your head up; few people can sell construction. It takes a special combination of knowledge, the hard hat in one hand and the calling card in the other.

I've only discussed the important areas of selling that apply to construction sales. Other segments have their special do's and don'ts. For example, the factory rep calling on his accounts on a regular schedule has certain rules he must follow, since even after he has opened the new account, he has to keep reselling every call. The real estate salesperson showing houses may be primarily selling to a particular sex, which also requires special knowledge. There are many other aspects of the selling profession as a whole, but we are only interested in the significant information that will make for successful construction marketing.

In closing this chapter, I want to point out that not everyone will be comfortable in this type of selling, and some will be comfortable only in certain parts—for example, the person who does not like cold calling may be first rate when in his own conference room. But please take the time to give it a good shot. Who knows—you may uncover hidden talents.

CHAPTER 3

CULTIVATING NEW BUSINESS PERSPECTIVES

You as a general contractor now have some idea of what has to be done to meet the prospect and, just as important, keep the lines of communication open. When this communication starts to take place, the general contractor begins to find that there are new business attitudes he must seriously consider using because they have a direct bearing on successfully closing the contract with the prospect.

Most contractors feel comfortable in the bid market. They know the market, and it's only natural to be comfortable with what you know and understand. Also, I'm sure every general contractor who has been in business for a while has negotiated work. Most of the time the negotiations you've made have been after you were low bidder and the owner wanted to cut the price even lower. Or you negotiated with an owner for a second project after completing the first job in a satisfactory manner. It could be a building that came from a friend or maybe the friend of a mutual friend of yours. In any case, all these jobs have one thing in common: they were presold!

That's right; you negotiated the contracts *after* the owner was sold on your doing his work. There is one whale of a difference between that kind of negotiation and the ones in which you have to sell the owner on letting you do the work, and then negotiate the contract. The new business attitudes are necessary if you want to operate in this area of design-build when the job has not been presold.

Taking the time to read this book shows that you've already decided to look into and maybe use the first attitude on the list, which is to go out actively and ask for the business, to consider marketing your services just like any other merchant. I'm not talking about sweeping changes, but I am talking about spending some money. So please read this chapter with an open mind; don't form any iron-bound opinions until you finish the chapter, or, better yet, the book. These ideas and guidelines are important to backing up the outside selling effort.

GIVING OUT INFORMATION

The general contractor is in business to make money. The one thing he does not do is give something away. Right? Not really, if he wants to sell jobs.

When bidding, the general contractor does not mind spending the time and money to put together a bid, and, believe me, I know how much of both will go into a bid. Then you, the contractor, go to the opening and sweat buckets. You're low bidder, and that's great. The only problem that nags at you is the money you left on the table.

The contractor goes through all these operations gladly because he's reasonably sure the project will be built. If the bids come in too high, then the low bidder still has a chance to negotiate the (presold) contract. So, all in all, bidding is worth the effort.

How would you like to spend some time working up prices, then give them to a prospect who doesn't have a clear idea if he is going to build? On top of that, he's not sure that what you priced is exactly what he wants. Can you handle this situation?

Please remember that you must keep an open mind until you finish this chapter. Keep in mind the question I've just asked as we press on.

Bidding is not something a contractor thought up; it had to be devised by an owner. And who can blame the owners? They want the lowest price they can get. The trouble with this system is that you, the general contractor, are always forced to tell the owner your lowest price. That's exactly why it's possible to leave money on the table; and I don't care who says anything to the contrary, it hurts to think you threw away a pile of cash. Then there's the bid you get that puts work on the books for the coming months, but very little profit. What's the percentage in it for you if you're just swapping dollars?

The shape of the bidding market is most likely the very reason that made you decide to give this book serious consideration. Remember the word *empathy*? (Look back if you're not sure.) The next thing we're going to do is empathize, to put you in the prospect's place. You're standing behind your counter when in walks this fellow, who looks around and walks up to you. He doesn't look like a customer or like the type of salesperson who calls on you. Your curiosity is aroused. Who is he? The man puts out his hand and introduces himself, then asks you if he can have a minute of your time.

Business is slow, so you tell him to go right ahead. You are still wondering what he wants. Then you find out. "Mr. X, I understand you are planning to open a new branch in the west end of town, and I'd like to talk to you about building the facility for you."

Well, he sure doesn't beat around the bush, you think.

"The best answer I can give you right now is yes and no. That's not much of an answer. You see, I have an option on the site, but I'm not sure I can swing it," you explain.

"Maybe I can help you decide," the construction salesperson adds.

"I'd appreciate any help I can get right now. My real problem is getting some kind of handle on the cost of the building. Then there's the site work, and my real estate agent has already told me there is a drainage problem that will have to be solved. I really can't make a decision until I get some idea of how much money will be needed. With that number I can then work out the projected business and know if I'll be able to cover the mortgage and overhead."

"How about giving me some of the details and letting me whip up some budgets for you to use?"

"Say, that'd be great; but wait a minute! What's this going to cost me?" you ask.

"Not a thing, Mr. X, except a fair chance to do the job. Now I'm not talking about a detailed quote, but a simple budget figure that won't take much of my time. Really it's the only way to get the ball rolling. You need the figures to decide, and if you decide to build, then I have a good project to try to put to contract."

Let's look at our little scenario from another viewpoint, one in which you're still Mr. X.

There's no change in our make-believe situation until after you have told the construction salesperson that you're not sure whether you will be able to build or not.

"Do you have any plans, Mr. X?"

"No, I don't," you answer, "but I do have some sketches," and you turn and hunt in your desk drawer until you find a line drawing done on a large brown wrapper. As you smooth out the paper on the countertop you think, maybe this man can get me on my way; I'm sure he'll be able to give me some ballpark figures.

"Is this all you have, Mr. X?"

"I'm afraid so. I didn't want to go to an architect until I was sure I could build."

"How about a site plan?"

"No." You feel that sinking feeling start; you know you'll get no help here.

"Mr. X. I'm sorry, but unless you have plans and specs, there's no way I can take the time to give you a price. After all, you're not even sure you're going to build."

What are your reactions from Mr. X's viewpoint concerning these two interviews? The last one shows how most contractors probably would react; and you as Mr. X would write that salesperson off and toss his card in the trash can. The first interview would offer you hope and a chance to get some information that could be very important to your pocketbook. You would keep the man's card and eagerly wait for the next few days to pass until you heard from him.

Okay, now stay in the guise of Mr. X, and we'll have the second interview. The salesperson has called and set up an appointment at your convenience. You picked late afternoon because business is slow then. The man (who seems like a real nice fellow) shows up on time. After the amenities, he gets right down to business.

"Mr. X, the complete building shouldn't cost over $260,000. Now that is everything: plans, plumbing, lights, pegboard on the walls—in short, a turn-key job."

"That's a little more than I'd hoped for, Mr. Doe [you had sense enough to get his name off his card before he showed up]. I was hoping for about $240,000."

"Well, there's a possibility it can be built for that figure, but please bear in mind this is a budget price and everything is estimated on the high side. I'm not helping you one bit in your decision if I give you a low price and you're knocked over with increases after you are locked in. Best to be pleasantly surprised with lower prices from a not-to-exceed building cost."

"That makes sense, but how about the site?" you ask.

"I must say we had to hedge a little on the site work. The least will be $35,000 and the maximum can be $45,000."

"Ten thousand dollars. That's a lot of difference," you point out.

"It sure is, Mr. X, but that bad drainage is the reason. There are various ways to solve the problem, and it will take some homework at city hall to see which way the city will allow us to go. [Note the *us* the salesperson has just used. It's his first reference to getting you to thinking about the two of you putting this deal together. He's now started making you a little dependent on him.] Time is the problem, and I feel you should use the high figure added to your building cost to give you a good budget project cost. I would like to add that my firm has some cost-saving ideas that we hope will appeal to you if you decide to push on." (This is a good, low-profile sell; he's not pushing you; he's letting you set the pace, make the decisions to build, also offering help in the vital area of getting the cost down.)

Though the figures are higher than expected, you are pleasantly surprised at the interest shown. "This is just what I need. The numbers aren't completely out of line. Now all I have to do is convince the bank."

"I can give you some help there, Mr. X," the salesperson tells you.

"How's that?" you ask.

"I'll give you a letter on company stationery outlining the project in general terms and stating the cost."

"Say, that'll be great. It'll make me look better down at the bank if they know my figures are from a contractor and not snatched out of thin air."

As Mr. X, how would you respond to the two approaches? I'm sure you would be all fired up by the salesperson who helped you start your project on its way. Suddenly there is somebody to light the way for you, to run interference, so there is an excellent chance you'll be able to build that new branch.

Let's stop pretending we're Mr. X and look at what our aggressive construction salesperson did out of the ordinary. The answer is simple—he gave away information. A better word is *invested*. Our man invested a couple of hours in time and some of his expert knowledge in Mr. X in hopes that his investments would pay off in the form of a construction contract.

You may point out that at least you know the bid job will be constructed, whereas the negotiated job at the beginning may still be a toss-up, and that's the fact that does not appeal to you. I have two answers to this reaction. One, I firmly believe that the higher the risk in investing, the larger the returns, as money-lending firms point out. You can deposit your money in a savings and loan, or invest in real estate mortgages directly and make a larger return; but you have a higher risk factor with the latter, hence more return. The same is true for construction selling; and, after all, isn't more money per job the reason you want a piece of the negotiated construction market? The second reason is percentages. All contractors deal in the percentages of jobs bid against jobs contracted.

Selling is a percentage business also, so think about this: On a bidder's list of ten, you have a 10% chance to get a job that you are reasonably sure will be built. On a negotiated prospect where our Mr. X is not certain he will build, you know the odds are 50% he *will* build; and if you are the only salesperson he is talking to, then you have a 50% chance of doing a job. Personally I prefer 50% to 10%. I know these percentages can change a hundred times depending on the circumstances, but, nevertheless, the contract-closing percentages for negotiated jobs are not as low as you may imagine. At the same time, if the negotiated job is going to be constructed, the number of contractors will be more like two or three trying to sell it instead of eight or ten as on the bid market. Still better odds in my book!

Providing information is one of the most important services with which a contractor can supply a prospect when the parties have a sell-and-buy relationship. It's strictly an investment and should be recognized as such by the general contractor, just as the time and effort to put together a bid represents an investment.

This investment will take the form of a budget estimate in most cases. It does not have to be a detailed quote, but only a ballpark figure based on your knowledge of construction costs. It's not too difficult to use square-foot prices.

Drawings

Usually a simple line drawing showing the basic requirements of the prospect is the starting point. When meeting with the prospect, the salesperson should know how to dig out information (details on this subject will be covered in later chapters) and put everything on the line drawing. This drawing doesn't have to be elaborate. I prefer to use 1/8″ grid in a handy size, say 12″ × 20″. This allows me to have everything to scale, which makes for easier and more accurate budgets. There is also a small bonus—it looks good for the prospect to see you using professional material. Remember, you sell your competency every chance you have; never let up on this point.

Not only does the contractor have to accept this new attitude of investing his time and money in the prospect but he must realize that the budget estimate he prepares will probably be the first of several. The first price is only the starting point. The prospect, upon receiving the first estimate, then begins to take an objective look at what he's trying to do. That's where the changes start, and these changes will require price changes. So be prepared to look at a prospect's job more than once; and, frankly, this helps the salesperson with his selling. The more you can see the prospect, the more chances you have to sell him on yourself and your company.

I know that some skeptics are glad to hear that there is a limit to the amount of money and time that should be invested in the prospect. There will be prospects who keep changing; they never have any idea what they want and will not hesitate to grind up your time for all the information they can obtain. (Incidentally, you don't have to worry about this type becoming embarrassed about using your time and services!)

It's difficult to tell at the beginning of negotiations exactly what type of businessperson you are dealing with, so I've developed a rough guideline that does a good job without letting you get in too deep with a prospect who is just using you: After determining the requirement, I work up the estimate; and then I will make two changes if they are minor and require very little time. I only make one major revision after the first one. In most cases this method will provide the prospect with all the necessary information to determine if he will build; or, if that has already been decided, then what he plans to build.

Please don't take this policy as a contradiction of what I mentioned in the previous chapter about prospect embarrassment and telling the prospect to call on you at any time for changes. The pricing guidelines I've just described are guidelines—nothing more—that I try to follow when actually putting out time and money for prices. I've found that the changes made by the prospect can easily be added to or subtracted from these figures, which allows the prospect to make changes without my having to reprice the entire job.

The next step after running through the price exercises is for the prospect to make a commitment to you to do the project so that the final details can be worked out and priced. There will be exceptions, and it's purely a judgment call on the salesperson's part. Don't hesitate to do whatever is necessary within reason to provide the prospect with what he needs if you feel he is making an honest effort and is not just manipulating you.

There is another segment of the budget estimate to discuss, but before that I want to remind you that you're supposed to keep an open mind. Okay, here goes. It's pretty hard to talk to Mr. X about his new project without some sort of drawing. I know what some of you are thinking. First, I tell you to give away prices and now I add plans! But before you toss the book on the desk, let me explain.

When you price a job to bid, an architect has already done all the work with the owner. It's been put on paper and given to the contractor to bid. With the design-build job the contractor is doing the preliminary work, some before the contract, with the bulk of it after the contract is signed. The one big difference for the prospect is that he has a true handle on the price when he deals with the contractor on a negotiated basis. At the same time, Mr. X cannot be expected to commit from price alone. Would you? I know I wouldn't. I would want to look at something!

The type of plan I'm talking about is a simple line drawing floor plan, and maybe a front elevation if it's a retail store, to show how the front will look. That's all I'm suggesting you use—not full plans and specifications prepared by an architect.

I know for a fact that some of my competitors go so far as to have the drawings blueprinted. I'm told they feel this looks better, and I'm sure they're right; but I think that's going a little too far until you've been told you have the job. Also, most contractors don't have a blueprint machine, so sending them out can run into money over a period of time.

Sub-Contractors

It's difficult to have up-to-the-minute prices in all the various trade areas; also, certain types of businesses will have special requirements that should be estimated by the particular trade involved. No matter what the reason is, whether price only or expert knowledge, trades will have to be called in to help with the budget estimate. It's important to have a working relationship with subcontractors who are willing to invest their time and money in your prospect. Their position is even less secure than the contractor's because the sub is not dealing directly with Mr. X.

My experience has been that the electrical as well as the mechanical trades (heating, air conditioning, and plumbing) are the ones I always touch base with when working up a budget. Many times the sub will have to provide a preliminary design in order to arrive at his price; so the line drawings that are used to sell the prospect also become the basis for the sub to budget his work.

There's nothing wrong with asking for prices from more than one sub in the same trade. The work may be complicated, and you need a good handle yourself before you give Mr. X his. The first budget from a sub may be too high; to use his price would raise your price higher than you feel will sell, so you look around for other prices in that particular trade.

I feel very strongly about the honesty and ethics involved in dealing with a subcontractor on a design-build project. If you use his price and sign up the prospect, then the sub should get the contract to do the sub work. I don't believe in trying to shop more profit into the job by putting that sub's work on the market; it's not fair to the sub because his budget price will have some money for the unknown hooker than can burn him—just as you should have in the budget you pass out to the prospect. It's very easy to shop around after the job is signed up and architectural plans are prepared because then the sub knows exactly what to price. I'm not saying for you to let the sub's budget price become his contract price. The sub will have to prepare a contract price from the final plans, and in most cases it should be below his budget because there are definite plans from which to work. But he should be the only one pricing from the final plans, since he should be locked in to you as you are locked in to the prospect.

In the case where you know what you're doing, ballpark a trade price, and sell the job, then you can obtain sub prices to suit yourself; but please remember, if you use a person's time and money, then you should try to work with him. Besides, it makes good business sense not to whipsaw subcontractors about. They can be a good source of leads. This was brought home to me recently by an electrical contractor who had done work for me. He and his wife saw my wife and me one evening in a local restaurant and came over to talk. While our wives were chatting, my friend told me about a lead. One month later I had the contract to build a 14,000-square-foot building. Needless to say, he did the electrical work.

I touched briefly on the final contract price while discussing the subcontractor, and I would like to elaborate a little on the subject. The budget price is used to sell the job; it's a tool, nothing more. It's supposed to enable the prospect to decide that you will be his contractor and also whether he will build. After this is accomplished, you, the contractor, have to get busy and have working drawings prepared in order to estimate the job correctly and present Mr. X with the contract.

Most of the time these will be the steps followed: requirements, budget estimate, decision by prospect, architectural plans, contract price. Once in a while the budget and contract prices will be the same, if the job is on the small side and the contractor has up-to-date prices he can use. My experience has been that this happens mostly with metal preengineered buildings with very little or no interior work. With anything else I never take the slightest chance—prices are checked!

ADVERTISING

Stay with me, keep that open mind, and I promise nothing will be as bad as giving away information and drawings. By going out to sell construction, the general contractor is now a merchant with wares to sell. He really is no different from the car dealer or beer distributor. You are going to have to start looking at advertising in a new light. I'm not talking about the special for the week as seen in the local newspaper.

Job Signs

Take job signs, for instance. Where else do you get the opportunity to put up a sign on the side of the road and tell the world what you do? Yet it seems to me that only the large projects get a decent job sign, and usually everyone who is anyone connected with the job has his or her name on it. What about the average job, say a 5,000-square-foot retail store? What's out front? From my observations it will be a 2-foot × 3-foot sign that shows its age. A prospect has to get out of his car, jump a ditch, and push aside some weeds to read the phone number. Well, I'm willing to bet that few do.

Granted that many general contractors are fully aware of

company exposure from job signs and put up nice large ones. That's great, but how well are they taken care of? They look fine at first; then the weather makes them show their age, or maybe one of our Monday-morning truck drivers backs into the sign, and for the rest of the job it's not even straight. Job signs are the best free advertisement you have, so it's important to use a large, well-done sign, and then keep it looking good throughout the entire job—and don't take it down until the very last minute. I've had owners ask me in a joking but partly serious way when I was going to remove my sign! So use them: big, well-done, and looking good.

Along with job signs go construction trailers, and you may use the side for your job sign. This is fine as long as the sign looks up-to-date. However, I prefer the job sign because it stays longer. You can put it up long before you move the trailer onto the site and leave it after the trailer is gone.

In Chapter 8, I present details of working with the owner during construction, but here is the very first step, which I'm bringing up now because it goes hand-in-glove with job signs. When you price the job, add a few dollars to have an owner sign painted and mounted right alongside yours. The first weekend after signs are up, the owner will probably spend all day Sunday riding by to see:

New Home **of** **Mr. X's Tire Service**

You've put him right out in front. Mr. X is moving up and he'll love it.

While we're on this subject, think about the location. Start out with the sign in the most advantageous place, even if it has to be moved later. A beautiful sign is worthless if not seen.

Yellow Pages

The next pertinent subject is using the Yellow Pages. Don't only be listed under general contractors; have a separate block that tells all you do in design-build construction.

Take a good look in your Yellow Pages, and check more than one section. For example, under "Buildings" the metal building dealers all have blocks advertising their building name and the services that they can offer the prospect. Let me add that if you are seriously thinking of marketing construction or have already started and don't have a metal building line, then you should give it thoughtful attention. It's a good business because you have something additional to sell, and the versatility as well as the economic advantages are important when negotiating the design-build commercial project.

People use the Yellow Pages. It's one form of advertising that the contractor will be able to monitor by logging the incoming calls. There is another big plus that is covered in detail in Chapter 5: the contractor will know the prospect is at least interested enough to call; a prospect who has already shown his interest is an excellent prospect.

Newspaper Advertising

Ever thought about advertising in the newspaper? You may want to look into it. Do some homework with your local newspapers, and get costs, sizes of ads, and the population you will be covering. Although a newspaper ad will get to a lot of people, only a small percentage will be thinking about building a new facility, but it only takes one to make the ad worthwhile.

There is one particular type of newspaper ad that is really more P.R. than advertising, but it does let people know what you do. This is the ad that congratulates Mr. X on opening his new facility. This ad may be a quarter of a page or even larger. Usually the general contractor can get the subs and suppliers to chip in, and in return their company name also appears in the ad. Maybe some of you have already been involved with this type of advertising. The owner is pleased as punch to see the splash in the paper, and it is good business because a satisfied owner can be your best sales asset at certain times. Admittedly, it's difficult to determine just how successful this kind of ad is. It's a lot like trying to assess the worth of job signs. You know they help, but you're not sure how much.

So take a look at your advertising. Bring in some professional help if you feel you need it. Just remember, when you have something to sell, spread the word.

ADVISER CAPACITY

The most amazing fact about prospects is their ignorance of the financial side of commerce outside their particular field. When you are bidding a job, the plans are provided for you to use. You assume, most of the time correctly, that the money has been lined up and all the owner needs is a price and he's ready to go. In other words, most of the messy little details are handled for you.

Design-build is the opposite; you are going to have to help advise the prospect through these areas. Details concerning construction loans based on permanent money take-outs will have to be explained in detail.

You will have to make sure the prospect looks good when he approaches the bank. I don't mean a clean shirt, though I have worked with some prospects where it wouldn't have hurt. I'm talking about a package of information to be given to the bank people that tells them Mr. X has his act together and is providing them with contractor-backed information about the new facility he wants to build. Nothing does the job like a drawing when those conservative bankers are looking over the prospect's application.

I provide what I like to call a bank package for the prospect to present to the loan people when the job has progressed to the point where Mr. X wants me to do the work and will sign the contract contingent upon obtaining his mortgage. At this time during the negotiations, it's in your best interest to do everything possible to help the prospect secure the money.

My bank package consists of two to three typed pages outlining the scope of the project as well as general specifications plus a floor plan and elevations. I have the drawings prepared by a draftsperson and then blueprinted. I have three, four, or five copies put together in a decent-looking folder for the prospect to pass out at the banks. I've found that it helps to have several copies for the convenience of the loan people at their meetings.

Along with advising the prospect on money, you'll be called on to answer questions about zoning, setbacks, sewer tap fees, and so on. It's possible the prospect cannot build what he wants where he wants, and you'll have to explain why and then help with the alternatives.

Before you say that you just don't have the time for all this sort of thing, think about how Mr. X is becoming more and more dependent on you. This is an essential part of construction sales—to make the prospect dependent on the salesperson.

As I have said, many businesspeople are ignorant of money matters outside of their special field. This makes you a natural to be called on if the proposed project has trouble and the prospect cannot, or does not, want to own the facility but nevertheless needs the new building for operations. In my experience almost any time the prospect has discussed leasing, it's been due to his needing the building and his capital also. Working out a lease package on a design-build situation can be very profitable to the general contractor, for if he has the means to obtain the site, either from the prospect who might already own it or by buying the site outright from a third party, the contractor will get a building contract plus the added value of investment property.

Most prospects really don't like the idea of becoming tenants and want to end up owning the facility at some time in the future, so they will be reluctant to enter into a lease agreement unless they have a buy-back option down the road, say no more than five years. If Mr. X wanted to rent all the time, he wouldn't be talking to you, but to a real estate agent with commercial property to rent.

Frankly, a lease design-build project is one in which everyone wins. The prospect gets the building as well as having it designed to suit him; he holds on to his operating capital while his business grows, and then when he's able, he can purchase the building. The contractor who signs a contract has invested in real estate that should have some tax advantages, and he should make a decent profit when he sells to Mr. X, since he sells at the fair market price at the time of sale.

Here's the hooker: nothing will grind up your time and money like working out the details for a lease, buy-back, design-build project—because you are doing it all! You're in charge of securing the construction loan, seeking permanent mortgage money, and handling design details with your new tenant, plans to banks, and all the dozens of problems that go with the building permits. When you roll the whole ball together, you'll wonder how it's possible to get it all done. But if you're willing to invest, then your return should be very attractive. Selling construction will require that you act as an adviser from time to time, but at the same time it helps make the prospect dependent on you, so don't hesitate to do it.

I've made the assumption that the person going out to sell construction is knowledgeable in commercial real estate to the point where he can talk to the prospect about it. If you're not comfortable with this subject, then I suggest you become comfortable. If not, you'll find yourself missing a real money-making job every so often. Also, it doesn't hurt to refresh the old mind in the subjects that might come up, such as setbacks, zoning, tap fees, and whether gas is available. Make it a point to stay abreast of commercial money interest rates and general developments in the construction money area. Competency in all these subjects makes you, the salesperson, look very good, and that's an integral part of selling.

REMAINING OBJECTIVE

I think objectivity is fairly easy to manage when the job has been bid off the exchange. Everything is cut and dried, spelled out right from the beginning. (Okay, I know that's the way it's supposed to be, but it isn't every time.) The general contractor has an arms-length association with the owner and usually the architect is also in the picture. This arms-length relation allows the general contractor to keep everything pertaining to the project on a businesslike and objective plane.

Negotiated work is the opposite; you are dealing with an individual before the contract, not after. Thus, the contractor is vulnerable at this stage. He's putting out time and money with no guarantee of a contract. All you have to go on are the other person's words and your own judgment.

Let me give you a very good example: when I was working for a metal building contractor as a construction salesperson, the owner passed a lead on to me concerning a certain Mr. X who planned to build a new automobile dealership. Needless to say, I went after the prospect hot and heavy. The car dealer had already hired an architect and put out price feelers using his plans. He was just testing the market, he told me, before he went out to bid. I can appreciate this; he wanted to get a budget to use when he talked to the bank.

The tentative prices he got rocked him back on his heels. No way could he get that kind of money, he told me. He was in a holding pattern. He needed some help to move the project along, and I went to work selling. I sold myself, then my company, and went into saving him money to bring the construction costs down to near his budget, which was based on this borrowing power.

I put a great deal of time into the project. Fortunately we had the plans to use, so my company wasn't spending money directly; we were spending it on my time and personnel in the office, however. After two weeks of work and three meetings with Mr. X, I recommended that the owner go back to the architect and have some large revisions made, which would offer savings from the construction standpoint. I told Mr. X I could meet his budget with the new changes.

Mr. X was delighted with the proposal and wanted a meeting as soon as possible to get the project under way. This conversation took place on a Wednesday, and I wanted to have the meeting with Mr. X the next day, but he was planning to be out of town Thursday and Friday. He set the date for the following Monday afternoon. (I didn't like the long time interval for reasons I'll elaborate on in Chapter 6.)

I showed up on time for what I thought was going to be a signing-up meeting. It didn't work out that way; it seems Mr. X had run into a contractor friend over the weekend who wanted to give him a proposal also. Being a friend, Mr. X agreed, and in good faith he couldn't sign with us and let his friend spin his wheels. He had no choice but to wait until the other price came in, which would be in two weeks. I left; there was absolutely nothing else that I could do. I stayed in touch with Mr. X for the next two weeks, and when the time period was up, Mr. X informed me that his architect had advised him to put the revised plans on the bid market.

I didn't like the turn of events one bit. Mr. X did assure me he wanted me to bid on the revised job. Several days later I picked up the plans, and when I unrolled them in the office, it was immediately apparent that the price was going to come in way over budget. It was going to be a waste of time; there was no way that Mr. X was going to spend the kind of money his revised plans represented. I decided not to do any more on the job. I was upset over the turn of events. I was going to sit back and wait.

Well, Mr. X was back to square one when the bid prices came in, just as I had predicted—way, way over budget. I honestly feel he was a little embarrassed to come back to me (remember what was discussed about prospect embarrassment), plus I think he was put out that my company hadn't bothered to bid. There was an excellent chance I could have salvaged the contract because I learned later that Mr. X had signed with a construction company that operated very much like my firm did.

Do you know why I missed that job? I was ticked off and wouldn't follow up. I didn't remain objective.

It's important when selling construction to remain objective about the prospect. Don't let little things get under your skin. I repeat, you're dealing with the prospect one-on-one. His way of doing business may be different from yours, so don't blow what may be a super job because of emotional reactions.

I'm definitely not saying that you should let yourself be used as a doormat. There will be times when your decision will rightly be "To hell with it." I'm simply saying that before you take that attitude, you should give the situation a good think-through and make your decision based on hard facts, not because you're angry.

Recently I made the choice to back away from a prospect. This particular Mr. X was interested in adding on to his business, so I called on him. He was the hardest person to track down. He had several businesses in the area and no one ever seemed to know where he was. I chased him for two weeks before I was able to see him. That happened at a lucky stop at one of his locations. We had a good meeting and I felt there were good possibilities for a contract.

I put together my prices and called back when I was asked to. Well, the same runaround started again, and after one day of trying to find Mr. X, I said, "Forget it." My decision was based on the fact I have other jobs to pursue and I have only so much time. I didn't become upset with Mr. X because he's a busy man and very successful. I looked at the situation from the dollars-and-cents angle only, and came to the conclusion that I couldn't keep my schedule in disarray so that I could make contact with Mr. X whenever I got lucky.

I'm the first to recognize that anyone in business will have unpleasant situations come up from time to time, and it's impossible to keep an objective attitude all the time. Okay, go ahead; make an emotional decision. I'll admit I've done so myself, but please don't make it a practice in construction selling.

I'll leave you with one parting shot before we move on to the next chapter: Remain objective about what has been discussed in these first three chapters. Even if not sold on an idea, consider giving it a try, but with an open mind that will let you remain objective. At the same time, be flexible and use what you find works for you. The first three chapters are general in nature and give the broad foundation that launches selling construction.

From here on, we'll be dealing with nitty-gritty details; and time after time, as these details are explained, we'll be referring back to topics that were covered in general in these first three chapters.

CHAPTER 4

LOCATING THE PROSPECTIVE OWNER

About now you're thinking, what good are all these guidelines, theories, and ideas going to do if there are no prospects to sell to?

This chapter will solve that problem for you. I'm sure everyone has heard the old adage, "Gold is where you find it"; the same applies to construction prospects. If there is one certain phase of construction sales where plain old hard work pays off every time, it's "digging up" the prospects. It's simply percentages—talking to enough people, looking deep enough in the right places, and keeping your eyes and ears open will always pay off.

Throughout Chapters 1, 2, and 3 the word *prospect* was used to denote the businessperson who was in the market for a new building. At this point I want to bring in a new term to call our Mr. X before he becomes an honest-to-goodness prospect: Mr. X is a *lead* before he becomes a prospect.

You have watched enough TV police programs to be familiar with the term *lead* and with how leads are used. The good construction salesperson will use leads just as the good police officer does. A lead can be a name or something you saw at someone's business address or a phone call from a friend; leads can come from anywhere. They are the one thing you cannot do without—no leads, no prospects, no contracts, no money. It's that simple. Until you qualify the lead and confirm that he or she is in the market for a new building or some kind of construction work, they're not a prospect.

The main thrust of this chapter is to show you how to locate the leads that can be turned into prospects. So please don't confuse the terms *lead* and *prospect*. To reiterate: A *lead* is someone you think might need a new facility. A *prospect* is someone you *know* is in the market for a new building.

It's important for the construction salesperson to have a prospect list and to constantly update it. To have a productive prospect list, the salesperson has to check out and follow up on a large number of leads. Many leads make a small prospect list that produces contracts. Obtaining and checking out leads is one of the most important functions of construction sales.

This chapter is about how to go about obtaining those all-important leads, so let's get right into it.

PERSONAL CONTACTS

This subject can cover a great variety of situations. You're at a party and a friend comes up and tells you about someone who's planning to build. Or it can be your banker or a salesperson calling on you who passes the word.

Some of the personal-contact leads are just plain luck. That friend you saw at the party might not have taken the trouble to call you at your office. There's nothing you can do about the leads you miss because somebody fails to tell you, but there is one positive action you can take—make sure everyone you associate with knows what you do. If as a contractor you're going into the negotiated design-build field, then let your friends know, both social friends and those in the business community.

There is a courtesy I always observe when a lead has been passed on to me personally. After I've checked the lead out, I contact the person who gave it to me and give him a status report. The report may be, sorry Mr. X isn't going to build, or he's thinking about it, but his lease has one more year to go. The important fact is you show the person that you took him seriously and you appreciate the lead. At the same time, your friend has the straight scoop about someone else in business, which might make him look good by being knowledgeable. This sometimes can be most important, since everybody likes to have inside information.

A fine source of personal leads can be clubs, either social or civic. If you don't belong to any, then I suggest you look into some. It's very important to know the right kind of people in the business community—the bankers, real estate brokers, lawyers, and loan company officials. They are the people located at or near the money, and everything in business is either going to or coming from the source of money.

Remember, it's strictly percentages; the more people you know and who know what you do, the more leads you will have. The old saying "It's not what you know but who you know" is very accurate. For example, a few years back I closed a contract that came from a neighbor. He told me that his employer had to move, his building had been purchased by the government, and why didn't I talk with him. So get the word around town; it'll pay off for you.

There are two particular groups in which I make it a point to have personal contacts, the first of which is commercial real estate brokers. Nothing will happen until Mr. X has a site, so the broker can be a good source of leads. At the same time, the broker may need your help because Mr. X won't buy the land until he has some idea of what the entire project will cost, and you'll be able to provide those numbers right up front. Develop a rapport in the commercial real estate business, and let the brokers know you stand ready to help them close a deal.

It's really a two-way street, and the contractor who's actively out hunting for business will turn over leads who are thinking about building but don't have their site. Thus, you are able to pass a lead on to a commercial broker. Now you're helping yourself, since when Mr. X obtains his site, he becomes a prime prospect.

There is one drawback when you are passing the lead, however: security (to be covered in detail in Chapter 6). Unless the broker can keep his mouth shut, don't use him. It's extremely important that the lead not get out on the street. If you dug the lead up, there's a good chance Mr. X may not be working with anyone else; and the fewer who know about Mr. X, the better your chances are for a contract. So work with a real estate broker who will work with you in every sense of the word, and make money for the two of you.

The second group that is a good source of leads is the subcontractors you work with, mainly the heating, air conditioning, electrical, and plumbing companies, since they will be doing different types of commercial work for the business community that is not new construction. A great deal of their work can be repair, maintenance, and renovation. These people will be going into a company and are in a position to notice and hear what plans the company may have. It can be a simple remark. The heating man asks why the new ductwork is not being extended to the rear section of the building, and the answer is that that section is to be replaced in the near future. That's all it takes, and the sub should be on the phone to you passing on what is obviously an excellent lead.

If you have the proper relationship with the sub, he really is passing the lead to you mainly to get himself a sure job—he knows that you're pretty good at negotiating contracts, and that if you get the job, he has the sub work in his trade. In other words, he's looking out for number one, but in the process he lets you do the same.

Remember in Chapter 2 we talked about aggressive selling. Well, there's more to aggressive selling than looking the prospect in the eye and telling him you want to do his work. You have to be aggressive in other areas, and that means taking the initiative. Don't sit around and wait for those people who can furnish leads from the business community to give you a call. People are generally busy and it's very easy for someone with the best intentions simply to forget. So give them a call every so often and ask if they have anything that looks promising.

Learn to pick up information every chance you have, whenever you see people who can furnish leads. (I'm speaking of business friends. I feel everyone has his own manner in mixing business with pleasure. My personal feeling is that it's okay if the other party approaches you, and then you don't stay on the subject very long. If it's something that should be discussed in detail, then make an appointment for business hours, the same as any other professional would.) As I stated, learn to ferret out information from these business sources of leads. Ask some questions.

Let me illustrate. A real estate broker called one morning and asked if I could give him a budget price. He was trying to put together a lease deal with a hardware company that wanted to expand. The cost of the construction was needed to work out what the lease amount would be. Naturally I agreed and turned the price over three days later. Two weeks went by; then I ran into the broker and asked how the negotiations were progressing. He told me he'd lost out to another group who'd also been making a proposal. There was a time when I'd have told my broker friend I was very sorry to hear he had lost out and walked off. But I've since learned better. When he told me another group had the lease, I had a hot lead to be pursued immediately. I casually asked who was the other group. The answer was some lawyer, and he thought his name was this or maybe that.

I thanked him and went straight to my office and called my attorney. He gave me two names of who he thought it might be, and on the first call I hit pay dirt. I went to work and presented my proposal several days later. The story doesn't have a happy ending, though; I didn't get the job. Nevertheless, the pertinent point is the manner in which the lead was "dug up."

Remember to *dig* (I know I've overworked that word, but it's the best word to convey the meaning I wish to put across); practice pulling out every scrap of information when there's a possible lead involved. People know more than they realize.

REFERRALS

Referrals are different from personal contacts, in that they come from people you've done work for in the past. A referral can also come from another contractor who can, but does not want to, work with the prospect for whatever reason. This is an extremely valuable source of business because you know the lead is interested in doing something if he goes to the trouble of talking to an owner and then contacting you. The problem with referrals is the time lag involved before they begin to pay dividends. You might wonder how this can be, since you have been contracting for years and have plenty of owners out there. But think: What kind of owner? Is it one who says, when asked by Mr. X about his contractor, "Good outfit, but you're going to need your plans first before they'll talk to you"? Or will the owner say, "Good outfit. They can take your project right from the start and do the whole thing for you, a complete turn-key job"? In my opinion the only way to get the second recommendation is to be in that type of design-build work.

Most bid jobs have rather long-range project owner relations, and when the problems begin to pop up, the contractor retreats back to the plans and specs and makes little effort to work with the owner. This has to color the owner's thinking about the contractor, so why should he pass out glowing reports about how great you are?

You're going to have to build your turn-key job reputation, and that takes time. It'll be a while before the referrals start coming in, but when they do, it'll be the source of leads with the highest percentage of contracts. The results can be outstanding.

ADVERTISING

I've already touched on this topic in Chapter 3, but let's look a little deeper.

Yellow Pages

The Yellow Pages is one of your more successful advertising methods. I find myself turning to them all the time. It's possible to keep tabs on Yellow Pages' results by the incoming calls, but you're going to have to instruct the person answering the phone to ask the caller nicely if his call originated with the phone book. Study the ads that your competition has and work up something for yourself. I imagine your local phone company can be of some help.

Incoming phone inquiries are very good leads because the people are coming to you, which expresses interest on their part. At the same time, you or whoever talks to the incoming caller will be able to secure necessary information, and this will save the salesperson time.

If your office is like most, you have someone who handles the phone, and often there'll be no one to talk details with the caller except for the secretary. In this case, train him or her to ask the proper questions so that the salesperson can do some homework before returning the call. Many times the lead is no lead at all, and a smart secretary can again save time.

It's important to find out exactly why the caller is calling. If you do only metal prefab buildings and the caller wants a block addition put on a shopping center, then your secretary should be able to handle the situation and hopefully refer the caller to another construction firm. If the caller turns out to be a bona fide lead, then your office help should try to determine the building size and just how the building will be used. The salesperson can have his mind in gear when he calls back, and that's important in helping the lead start to develop a good impression.

A word here about calling back. *Do it!* And promptly. That's the first good impression the caller will form. I'm absolutely amazed at the number of businesspeople who don't take returning phone calls seriously.

Direct Mail

This is another area of advertising that can be used. Like newspaper ads, it covers a lot of ground, but the advantage is direct mail will go to people that may be in the market for a new facility, be it now or in the future. Also note the fact that the recipient responding to the direct mail shows interest, and this is important. Granted some of this interest is simply getting information for whatever the reason, but still it only takes a few bona fide leads to make the mailing worth the effort.

Teaming up with suppliers that promote direct mail is a good method to follow. As a general contractor you may have already been involved with working with a supplier and feel comfortable in this area. If so, great; if not, check it out. The metal building industry is very keen on direct mail campaigns, which is another reason to consider a metal building line as part of your business.

Calling Card

I consider the calling card a very special form of advertising, and for it to be effective it must inform the reader of exactly what you do. I have seen who knows how many cards that listed the company name, the person's name and title, fax, phone number, and address—but not one word describing what the firm does. I guess if the word *construction* is in the company name then that's all the job description that is needed. However, I feel you should take full advantage of every opportunity to tell what you do, so under the company name outline any particular area of construction in which you specialize. For example, my card stresses commercial construction and metal buildings under the company name. I carry this two steps further. On my company letterheads and envelopes the same areas of specialization are listed.

My next recommendation concerning calling cards is always have one handy no matter what you are doing. It makes no difference if the group you are with is social or business, and no matter the size. Be it a couple of people you are having lunch with or a huge wedding reception, always have a card handy, because if the conversation gets into what you do, it's a simple matter to pass over a card, or better yet, to exchange cards, especially if the card you get is a good lead.

I find the card very handy. Whenever I stop in a business place that has a form to fill out, I put the card on the counter, which makes it real easy for the clerk. While I do this mainly for convenience, you never know when something positive may be uncovered. I have a good example that recently happened. For recreation, my family and I enjoy boating on the Chesapeake Bay. The first thing you do when you visit a marina is check in at the office and take care of the paperwork. Again to help keep things simple, I use a calling card with the boat's name written across the bottom. Most of the time the card is not returned, and in one case, after I had checked in and was turning to leave, the owner came into the front office, noticed the card, picked it up out of curiosity, I am sure, glanced at it, then asked if it was mine. I answered yes.

He then invited me into his private office and explained he had some serious leaking problems with two of his metal buildings, so since my card stressed metal buildings, would I please have a look at the problems. I did and made recommendations to solve the problems. He then asked me if I could do the work. It was tempting, but a little too far from my working area, and frankly I had all I could do at the time, so I declined. But I did tell him I had some contacts and someone would be calling him.

When I got back home I called my metal building salesperson, who serviced the area of the marina as well as my area, and passed the lead on to him to give to his local dealer for follow-up. So when it comes to calling cards, always have them handy.

TRENDS IN SPECIFIC BUSINESS AREAS

This subject will take a little more work and observation on the salesperson's part. Business trends of all types are taking place constantly. Some are quite obvious, and then again some won't be noticed unless you take the trouble to look. For example, in 1990–1993 when most of the nation was in a recession, the contractor and many other people found it hard going just to keep the doors open. Kindred souls were the automobile dealers. Can you remember when they would do anything to sell a car? While the car dealers were having rough times, as were many allied businesses in the automobile industry, auto parts stores were doing great. People were driving what they had longer; therefore, they needed parts. Also prospering were repair garages, which made money keeping the old car on the road, at the same time buying parts from the parts suppliers.

Suppose motorcycle sales have gone up astronomically over the past few years. There are many reasons why, which are not our concern, but the fact that sales are up makes cycle dealers a prime target for leads. At the same time you are doing your homework on a business trend, make it a point to learn what you can of a general nature about the business. Sometimes you can get help from an unknown source. We're talking about the cycle dealer, so consider this: I understand that most started out as second lines for car dealers, and the motorcycle manufacturers don't want their cycles marketed alongside cars—they want a completely separate facility for the bikes. Some dealers could be required to separate the two; in some cases it could be the car manufacturer asking for the split. Whichever, this kind of knowledge gives you a good source of leads that you probably wouldn't have pursued had you not taken the trouble to learn something of the industry in which you are interested.

The trend can be going on in any business area; all the salesperson has to do is dig around. Talk with people who deal with the whole spectrum of business: bankers, stock brokers, real estate brokers, lawyers, and mortgage loan officers. Talk with friends who are in other business fields. A good place to maintain good relations is city hall. Maybe there is some new regulation that will cause some special construction to take place. For example, you might discover that the health department has a new ruling about the storage of certain kinds of food, and the food distributor will become a prime area to be worked.

Trend leads do have the drawback of being somewhat long-range if the salesperson is on the ball. That is, he may call on Mr. X before he is really ready to expand. Mr. X knows his business is good; and if you caught him early, it'll take some time before he's ready to build or add on. Once in a while you'll catch a fast-moving trend lead, but my experience has been that they have to be nursed along for a period of time. How long? I can't say. We're dealing with people; therefore, there's no yardstick. I'll give it one quick swing based on past experience, though: 6 to 12 months.

I consider publications as a good source of trend information—the better news magazines, financial papers, and trade journals for other fields. I make it a habit to look through any trade magazines that are in the lead's waiting room, and you don't have to be in an office. Just about every business will have trade publications lying around somewhere. Look at them; never forget that leads are where you find them.

Environmental Construction Trends

To carry the business trends concept a step further, let's address environmental construction. As noted in Chapter 1, this area is beginning to really affect the construction industry and also fits in very nicely with the design-build concept.

While business trends are market driven, environmental construction is not. This trend is driven by regulations, be they local, state, or federal, that the business community must take into consideration, as well as the private sector.

Here is where the environmental professionals part of your design-build team can be a big asset. Stay in touch and work with them to see what particular areas can be pursued for business prospects.

When you start making environmental sales calls, you'll probably note the same thing I did, which is often the person you are speaking with is not up to date on the regulations, and he's learning from you there may be a big problem to take care of. The important thing is there is somebody now in the picture that can solve the problem.

REDEVELOPMENT PROJECTS

Redevelopment projects can be a veritable gold mine of leads. It seems that almost every city has a program to buy up old sections of town, bulldoze the buildings, and rebuild. The businesspeople in these areas are being forced to move, making them prime leads to be called on and worked. I have personally found that many of the older established distributor types of business are located in these older areas of town, giving the construction salesperson not only an excellent lead but one that appeals from the standpoint of both job size and establishing an easy rapport with the buyer.

The place to start is at the local redevelopment project office. Gather all the pertinent information possible, talk with officials, look at maps, and come away with anything they'll give you. The redevelopment officials will certainly not be able to recommend you to a prospect, but they should have no qualms about telling you about Mr. X down on Sixth Street who's dragging his feet about leaving. The officials may be good enough to pass on lists of names whom you can contact, which allows you to call first. One very important fact should be uppermost in your mind, though: the lead is faced with having to move. There may be a time element involved, but at some point he's going. Rest assured that you cannot find a better lead to work.

Now if you are aware of redevelopment projects in your area and are familiar with them, you may feel there's no need to waste your time talking to the officials. You'll walk right in and meet Mr. X. Don't do it; always check in with the office and learn all you can. After all, the area you can see being redeveloped may be a small part of a much larger whole that only shows up on some map in the local office. While you are at it, make it a point to stop by every so often and keep up-to-date on the changes.

RELOCATION PROJECTS

Relocation projects are similar to redevelopment projects but have one main difference: when you see the work taking place, chances are you're too late to sell a job. The relocating businessperson who has to move will have all the details behind him before he moves out and the bulldozers move in. Again these leads are prime ones because Mr. X is being forced to move. No question about it; he will build a building.

Relocation projects bring to mind right-of-ways for new highways or roads: This area offers very productive leads. At the same time, expand your thinking into other fields that might offer leads. How about business areas around a fast-expanding unit, say a college or a hospital complex? Military installations offer good leads, and I've just had this brought home to me in a very positive manner. The Langley Air Force Base enlarged the safety zone at the end of one of the main runways, and the government bought out three businesspeople. Two rebuilt, and I negotiated a contract with one for a 2,600-square-foot office building. Point of interest: The other owner had some construction contacts and acted as his own general contractor and rebuilt, so I didn't lose him to my competition.

You have to work at finding relocation leads. Watch the paper; my experience has been that relocation projects are advertised in the papers as well as being carried as a news item. The second way to discover what's going on is to find out the government agencies (federal, state, and local) that are concerned with relocation and then go see them from time to time.

Of course, you'll learn when you start looking for leads that one lead-gathering method will help with another; the lead will be the result of the combining of two or more methods. Say, for instance, a friend at a party happens to mention a relocation project—no names, just a lead for a lead, which provides the salesperson with a prime area to investigate.

I repeat: Leads from relocation projects are extremely productive because Mr. X has to move.

INDUSTRIAL COMMISSIONS

Most larger cities are very interested in bringing new industry to their locale; therefore, there should be an industrial commission that the construction salesperson can contact for leads. Many times the commission operates an industrial park so that ready sites can be provided to the prospect shopping around for a new plant location.

It will pay the salesperson to check out this area and become acquainted with the officials. I do want to point out one important fact: This type of lead is not as productive as the ones we have covered earlier, the reason being that the lead in the beginning is just looking around. He has no site and maybe is not even sure of his desired location.

Work with these leads only after you know that a site has been obtained. The out-of-town lookers are notorious for taking your time and information and then moving on. Don't ignore this source of leads, but keep both eyes open when you're working with this type.

REJECTED BIDS

I particularly enjoy working with this segment. I've had several good money-making jobs from the rejected-bid ranks. This source of leads is also an easy area to pursue. Just sit at your desk and look through the material from your bidder's exchange, and then pick up the phone. Rejected bids are not leads but prime prospects because the owner has everything together and is ready to go. Then, wham! He's knocked over by the prices and wonders what to do next.

Suddenly the construction salesperson appears on the scene and starts showing Mr. X how to lower the cost and hopefully build the facility without losing needed requirements. A personal experience illustrates exactly what I'm talking about. It happened several years ago while I was working as a construction salesperson. A local equipment company had plans prepared for a new facility; it went out to bid and was floored with prices $50,000 to $60,000 over budget. I came across this fact through one of its outside salespeople before it became public knowledge and immediately contacted the owner. I picked up a set of plans and my company studied them for a week with the aim of cutting construction costs without decreasing square footage. At the end of the week I met with the owner and told him we could not only lower costs but there was a good chance the final price could be under the original budget. The owner, who was one sharp individual, answered my presentation with two words: *Show me!* I explained that if he could spare the time I wanted to arrange a conference at my office so that the estimator and owner of my company could explain in detail plus have the figures handy if he wanted to think about other changes.

My reasons for setting up the meeting were twofold: first, I did want our estimator, who was a licensed structural engineer, to explain in detail what was proposed; and second, it's a good selling technique to show the prospect he's dealing with a first-rate outfit—and what better way than to let him see a busy, well-run office? We showed the owner that the structure he wanted to build was basically a metal building but not of the standard preengineered variety; also, the exterior panels were of a very expensive material. Could he afford that much money for appearance? No was the answer.

Our recommendation was to substitute a standard preengineered metal building with the manufacturer's standard exterior panels. This represented a substantial savings. Another big savings came from the parts department. The original plans had had this area air-conditioned because the owner had so instructed. No one had bothered to explain to him that this was a large volume of space to cool and it would be costly. Instead we recommended using a mechanical air-change system that would bring in outside air while exhausting the warmer air from inside the area. After all, the shop wasn't air-conditioned, and the parts department was attached to the shop area. Why hand a nice cold part to a mechanic that you are paying to cool?

There did have to be some provision for the parts people to be able to work in warm weather; hence the mechanical air-change system. All the mechanics had to do was open the huge door at each end of the shop for extra ventilation.

Another place where we saved money was the glass in the front office section. I recall that a special bronze glass was specified, but the front of the building had an overhanging mansard. Also, many large oak trees shaded the entire front. We suggested regular clear commercial glass, since the sun's heat wasn't a real problem.

We suggested quite a few more money-saving ideas, but they were all small, so I won't get into them now. Needless to say, the owner contracted with my company to build his new facility, incorporating the money-saving ideas. Our final price was indeed under his original budget figure and he didn't lose one square foot of area. It was also a profitable project for both the company and me personally. The ironic thing was that the lead didn't come from the bidder's exchange but from a personal contact, which proves that business is where you find it, or in some cases, where it finds you.

The point to remember about rejected-bid prospects is speed. You have to move fast. As I said earlier, most negotiated jobs take time, but the rejected bid is the exception. Our Mr. X is ready to go; everything has been worked out—site, plans, money in most cases; all he needs is some help getting the price back down to earth. Unless there's a special situation with a lead or prospect, I personally put the rejected bid right up front, with precedence over anything else I happen to be working on.

While on the subject of rejected bids, I made the assumption that the construction salesperson or someone else in the company has the ability and know-how to actually redesign a commercial project, with savings recognized from different construction techniques and types of materials. You will not have much of a chance to sell the rejected-bid prospect if you can't offer him some tangible evidence of how you're going to do what you say you're going to do. If you need help in this area, get it; don't try to bluff your way with the prospect. At some point it will show that you don't know what you are doing, and Mr. X will give you your walking papers.

Not all rejected-bid negotiated contracts will go as smoothly as the example I have just described. Sometimes the owner won't be able to get all his requirements. It may not be feasible to bring the cost down to what the owner is looking for without some compromise on his part. When this happens, I push the idea of "better something than nothing." I stress compromise now with expansion designed into the plans to pick up the dropped requirements in the future. When informed of the cost of items in the original job plans, the owner may find there are some he can do without. For example, when Mr. X starts moving up in the world, he tells himself he's going to do it right. So he instructs the architect that he wants a large office with adjoining bath. Then, almost as an afterthought, he adds a shower to the bath. Mr. X is a busy man and many times would find it convenient to change clothes before attending a club meeting in the evening. The architect says, "Sure thing," and draws in a 20′ × 20′ office with a ceramic-tile full bath.

If this job reaches the rejected list, the construction salesperson may have to ask Mr. X if that big office and bath are really necessary because they're costing so much money. Could he get along just as well with a 12′ × 12′ office with a half bath that doesn't have ceramic tile? My experience has been that the owner quickly cuts out what are obviously luxury features and starts working with the hard monetary facts so that he can build the project he wants so badly.

The luxury compromise doesn't really cause the owner very much concern; he's a businessperson and facts are facts when he's faced with the decision of whether the money spent will help him make money. On the other hand, a compromise that may interfere with his making money is something not to be taken lightly; it requires some thought.

The future owner may conclude that he's going to cut and compromise only so much. His next step will be to accept a higher budget and go after more money. Guess who will be standing right there with dependable, realistic prices plus any other help when Mr. X returns to his bankers? That's right—*you!*

The one thing the construction salesperson must always remember about rejected bids is to get the job off the bidder's market and then have Mr. X commit to him as quickly as possible. Never lose sight of this; the longer the project stays on the street, the better are the chances that something will happen to foul you up. Believe me, it can be some small insignificant point that blows the whole deal.

Remember, the rejected bid is not a lead but a prime prospect, because all the preliminary work has been done. He wants to build, so give the prospect top priority and move fast!

COLD CALLS

In this particular type of lead gathering, the construction salesperson must be able to think on his feet and observe details. Many salespeople don't enjoy making cold calls; they're uncomfortable going in when not expected. To be successful, the construction salesperson should adopt the same attitude the old prospectors had, that "the next rock I turn over could have the mother lode under it." Every door you walk through could have a half-million-dollar contract waiting.

When a construction salesperson makes that cold call, he's been able to observe the physical surroundings. Maybe he noticed material stored in the open or obviously incoming boxes left on the sidewalks; or it could be a garage almost covered up with cars to be worked on. In other words, the salesperson is able to observe that the businessowner might be in the market for a building, and armed with this knowledge, he walks right in and introduces himself.

Cold calls are a necessary source of leads that should not be overlooked. Admittedly, some construction salespeople regard them as a necessary evil. The physical act of cold calling is often combined with another lead source. For example, the lead may come from a personal contact, and then the cold call is used to see the lead. Granted, this method should be used only as a last resort. If you have the lead's name, it will be possible to call and arrange an appointment. In my experience, when I have combined cold calls with any other source of leads, it has been because I didn't know enough about the lead to feel comfortable calling ahead, but this rarely happens.

Maybe I should give you my definition of a cold call before you skip the rest of this section. A construction cold call is one in which, no matter what the reason, it has been determined that someone may be in the market for construction of some type—new, add-on, or repairs. Please be advised that I'm not suggesting you start at one end of the street and proceed to the other end, knocking on every door as you go. There's just not enough time; besides, that is a very unproductive use of your time.

I've worked out a very special use of the cold call, which you might try: I use cold calls as time-fillers during the day. I never set aside a certain time to make cold calls; as you well know, in the construction business too much happens on short notice that requires your attention.

For example, let's say I have a 3 P.M. appointment and I've just left a job I'd been checking at 2 P.M. That gives me an hour, which may be lost time, so I'll put that hour to good use. I'll visit a part of town that I haven't seen for a while and just look around. I simply cruise about observing the businesses, looking for the telltale signs of growing pains. When I spot a likely looking lead and there's enough time, I go right in and start asking questions. Sometimes, because of time constraints, I have to return to make the call, but the lead is investigated.

I try to make approximately five cold calls a week; some weeks I make more, and some I make fewer or none. The construction salesperson should always keep the cold call in the back of his mind and not only plan to make it, but do it! Cold calls require discipline. It is easy to put them off. I know; I do it sometimes. It takes only a weak excuse not to make the cold calls you know beyond a shadow of a doubt you should be making.

A construction cold call is a qualified call; you don't stumble around trusting blind luck. The successful salesperson knows like the back of his hand the signs that show when a business is growing. He's always looking about while driving, rather like a hunter after game. Common sense tells you what to look for, such as old buildings; materials stacked outdoors that really should be under cover (industrial firms display this telltale sign); jammed loading docks and truck parking areas; in the auto industry, many more cars surrounding a building than seems normal (this is a good indicator for auto garages and allied specialty shops such as radiator or body shops); trucks loading and unloading on the street through a front door with goods stacked on the sidewalk.

The details will differ for different types of businesses, and there may be special details to look for in certain areas, say in the rural business community as compared with the metropolitan segment. But one point will not change: there will be some sort of telltale sign of growing pains, and that's your tip-off to make the cold call. A little practice will go a long way here; and the beauty of it is that you can do it while going from place to place, which gives you maximum use of your time. Time is the one thing you'll never have enough of when you really get into selling commercial projects.

You may wonder why you should walk in on Mr. X. Why not jot down the name and address of the business and call? There is one good reason: the fast brush-off—and not because Mr. X is rude and hard to get along with. The brush-off I'm speaking of is the kind every salesperson gets if he's out there doing his job. Our Mr. X is busy; he's up to his knees in alligators when his secretary tells him there's a Mr. Somebody on the phone about a new building. Remember what we said before: the lead or prospect will always put his customers first and you second; so Mr. X gets on the line and tells you he doesn't have time to talk now, but maybe later. Keep calling and catching him busy, and suddenly you're a pest.

Well, you ask, what's the difference between this call and one you are following up from a personal contact? It's simple: most of the time with a personal contact call you have an entry, the mutual friend. When you get Mr. X on the phone and mention you're calling because Jack Smith suggested it, then Mr. X will be more inclined to give you a minute or two. You aren't a complete unknown; there is a line of contact between Mr. X and the strange voice on the phone. Take that line of contact out of the conversation and you're next to nothing; the only way to overcome this problem is to let Mr. X have a look at you. At the same time, you can look Mr. X over and try to learn all you possibly can. If Mr. X is busy, you can wait until he has a minute; you'll be able to alter your game plan to suit the situation and make yourself more acceptable. Most businesspeople will take the time to speak to someone who takes the trouble to call on them. So you're able to get to Mr. X and start qualifying him to see if he can be moved into the prospect column.

In the event that Mr. X doesn't have the time to see you, at least you have a slight entry if you prefer to call before you try to talk to Mr. X again. I know it's not much of an entry, but it's better than nothing.

Of course, the lead can be checked out sometimes without talking to the person in charge. If for some reason you aren't able to see Mr. X, talk with his employees—the second in command or even the secretary. Just a simple statement, such as "Mr. X has just signed a new five-year lease," or "He's taken on a new line and is really going to need some more space," will let you know how to spend your time on this lead.

The best way for a construction salesperson to save time is to find out immediately if the lead is going to build. Knowing when and with whom not to waste time is the surest manner in which to save time.

The cold call is a source of leads that has to be worked physically by the salesperson. You should be aware of one fact about cold calls. They usually are long-lead-time projects simply because Mr. X probably has done very little to nothing about the project he's thinking about. The exception is repair work; this usually moves quickly.

Also, the cold call can be put in the same category as direct mail. It takes a lot of contacts to generate a small list of leads. This is because you know nothing about it except what you can personally see; it's not like an incoming call from a referral, where the simple fact that Mr. X has called you expresses interest on his part. The redevelopment and relocation project leads both have owners who have to move. You even have some degree of interest on the part of the lead with personal contacts and referrals. But with the cold call, you have nothing until you go in and find out. Simple mathematics tells us it takes a large number of cold call leads to generate prospects that turn into contracts. The cold call contract takes more work as a whole, but it's still an excellent source of leads and should be used.

Now let's talk about a very special area of cold calls, which I'll refer to as *prequalified*. These are leads that offer a better chance of results because of a particular situation. I have a good example. A few years back I realized there are a lot of metal buildings starting to show their age. Therefore, the retrofit market was one that needed to be investigated in an aggressive manner.

In the course of moving about during the day, I took note of older metal buildings and then grouped them into areas. It's easy to spot the signs of age, that is, rust, ding marks, faded paint. It's amazing when you get into something like this how much you know and remember about your area of operation concerning buildings you did not build, which is a big help.

So after a few weeks I had a list of older metal buildings grouped in areas of convenience. From past experience, roof problems were going to be the main complaint on the older metal buildings. Knowing this, I waited until my location (which is coastal Virginia) had a good Northeaster off the ocean with howling winds and heavy rains. The very next day, as the sky cleared, I devoted a whole day to calling on the buildings on my list. What I was after were reroofing jobs. The results were two repair jobs, a good number of future leads to stay in contact with (some have since called me for repair work), and a reroofing job that led to other work. All in all a good day's work of cold calls.

Now back to my example. This firm was a high-tech machine shop doing work for the government, and leaks were more than a mere inconvenience. When I walked into the lobby I asked the receptionist to let me see someone in management about their roof leaks (assuming they had some). Their problems were so bad that I was talking to the top man in short order. In the course of the conversation I was asked how I happened to stop in, because the top man assumed at first that someone in plant engineering had called me. I told him the call was a "qualified cold call" looking for business. Frankly he seemed pleased someone thought enough of his business to seek it out. I got a contract to reroof part of the building, then 14 months later I built a 5,000-square-foot model shop, and 11 months after that a 10,000-square-foot addition to the main shop.

Learn to observe places of business, and then learn to notice changes that occur over a period of time. Always make it a point to keep returning to the business location you feel offers cold call leads, so that you might see the changes that take place over a period of time. Also, remember to get out there and diligently make cold calls!

The Secondary Cold Call

I know you're wondering just what this is. It's a very special use of the cold call, and the word *secondary* is used because the actual cold call is the result of another action.

I'll start with a direct question. Have you been in a hardware store, or maybe an auto parts store, or say a lumber yard or any type of business along these lines recently? That is, businesses that need fair to good-size storage space? If the answer is yes, did you seek out the owner or manager and ask if he had any type of expansion plans for whatever the reason?

In other words, you're buying something and when that's done you make a cold call. Don't laugh. It works, as the following example illustrates. A little over a year ago, one of the electric eyes on our kitchen stove broke. The next day I stopped in at a local appliance parts dealer and bought a new eye. Then I asked to see the owner or manager. In short order I was talking with the manager about expansion plans. I told him what I could do for him, and he was quick to point out he needed more storage space badly, and for me to contact the owner who's main office was located in another city 25 miles away. The local operation was one of four branches.

I called the owner and stressed the fact I was calling at the suggestion of his manager (please note I'm not just a voice on the phone, but someone known to one of his managers, which gave me an introduction). The owner explained to me that they indeed needed more storage space, but one thing stood in the way. The problem was that the land behind the store, while owned by him was not zoned for business, and he had no idea if it could be. I told him I had some experience in this area and I would be glad to help when he was ready. I said I would stay in contact in the coming months. (Notice I didn't say call me when you are ready. I wanted to be the one checking things out.)

Anyway, we worked out the zoning problem, and I built a 2,500-square-foot storage building behind this appliance business, and all because I made a cold call after buying the stove part. So make the secondary cold call when possible.

DISASTER LEADS

We've all heard the old adage, "It's an ill wind that doesn't blow somebody some good." Well, there have never been truer words with respect to construction salespeople and disasters involving commercial buildings.

Fire is by far the main disaster that strikes buildings, and then Mother Nature can get upset and unleash a tornado or hurricane. No matter what destroys or damages a structure, get yourself in the picture as soon as you can, considering the owner's personal feelings at what must be a trying time. Every once in a long while you'll encounter the fellow who smiled all the while the firefighters worked to put out the fire; the fire solved his problem. He takes the insurance money and builds a new facility where the old one stood.

Ninety-nine times out of a hundred it's imperative that the owner start a new building or repairs as soon as possible. Here's a good example, which I feel shows what the aggressive salesperson can accomplish when he goes after a prime disaster prospect. When I worked for a design-build construction company that employed another salesperson as well as me, my colleague was working with a man in the furniture business who wanted to open a new branch. The lead, by the way, was a personal contact, which came from a club meeting.

Anyway, this salesperson worked like crazy to sell the job. There was another club member after the job also, and the owner chose the other company's proposal over ours. I remember the reason—the other company's price was a couple of thousand dollars lower and the owner opted to save the money. That's what the owner expected, anyway; but well into the job it suddenly came to light that there was quite a lot of fill dirt needed and the cost wasn't included in the contract price. The end result was that the owner paid more than the price my firm quoted, since we had included the fill.

The owner was good enough later to admit he should have paid closer attention to details and signed with my colleague. Six weeks later, on a windy Saturday afternoon, the furniture man's main store burned to the ground. That night my colleague called offering to help in any way, and this time he did build the new building.

This is also an excellent example of remaining objective and not getting personally upset when you lose a job. You never know when that owner will become a prospect again!

Disaster leads can be good ones because the prospect will have to have a building from which to operate. Follow the news media and pay close attention when you hear or see news that may give you a lead.

BONUS TO EMPLOYEES: FINDER'S FEES

Make salespeople out of all your employees. Stress to them the importance of bringing every scrap of information that may be a lead to the attention of the proper person. Instruct your people to keep their ears and eyes open. It's positively amazing what they can bring in when they are on the lookout for leads.

You can give the employees the old college pep talk about how much the leads mean to the firm and what's good for the firm has to be good for them. When the first lead comes in from your bookkeeper and ends up in a big fat contract with a good margin of profit, you could give her congratulations and a big thank-you in front of the whole office staff. Months later you might still be trying to figure out why the hired hands haven't brought in any new leads since that last great one.

Well, nothing in this world is free. Your employees are paid to do a job, nothing more. Why should they go out of their way to put money in the boss's pocket? Their check arrives regardless of the leads. The answer is to make it worth their while. Give the bookkeeper a big thank-you and a check in front of the office staff, and notice how there'll always be a lead or two coming from the help. Nothing else, absolutely nothing, motivates like money, so use it to make money.

Don't be selective; let everyone who works for you know he or she has the same standing offer for a lead that ends in a contract. Here's the catch: the lead has to become a contract. This idea of a bonus to employees will work and provide the firm with a new source of leads because seldom will the employees associate with the same people that the owner does.

Another paid source of leads can be finder's fees. In this case you have an understanding with salespeople in other fields, say equipment salespeople, or insurance people, or maybe a commercial real estate broker, whom you will pay a set amount for a lead that turns into a contract. These people are in and out of many businesses and know what's going on. For example, why are certain businesspeople talking with the equipment salesperson about a new 10-ton bridge crane? Why are they talking with their insurance agent about the cost of their coverage if they happen to add 50 people to the shop payroll? Why are they talking to the real estate agent about buying the adjoining property? It would take a real dummy not to be able to see business expansion here.

Others have knowledge that you lack. Realize that you might get your lead as the result of a personal contact, and that's great. But personal contacts are personal; the person with the lead is doing you a favor by telling you—there is no incentive most of the time except for helping a nice guy or a friend. If you could only know the leads that were meant to be passed on to you, but fell through some crack for one reason or another, you would just be sick. For every one you get, I personally feel three or four get dropped. The answer could be a finder's fee. Nothing else motivates like money. But you can't go around offering money to your friends for leads. You can, however, offer it to people in business whom you see mainly during business hours.

I personally use the finder's fee offer very sparingly because I don't like the idea of being known for having to buy my leads. This really is a personal judgment situation for the construction salesperson. In other words, it's hard to tell someone how and when, if at all, to use the finder's fees. There may be accepted business practices in your area that can help you decide on their use.

My own guideline is that when someone brings me a deal, not a true lead, and it puts me in the know, then I'll build in the fee. For example, a person brings me a possible lead, such as that he knows about a new building, and he's trying to sell some part of the overall picture to Mr. X, and could he make a few bucks if he gets me in to see Mr. X. In other words, the lead-seller is shopping before the fact, and that's only good business—for me and for him. When I know what the situation is and have some measure of control, I will consider a finders fee.

ZONING CHANGES AND DEED TRANSFERS

Watch the paper and/or check with city hall for applications for changes of zoning to accommodate commercial enterprises. These are good leads because someone is planning to build and is going to considerable trouble to have the site rezoned.

Deed transfers are another good lead source. When Mr. X puts out hard cash for land, you know he's serious. Often deed transfers are recorded in the local papers. I'm sure it's different from locality to locality, so you'll have to find a way to keep updated on who buys what. Maybe your area has a publication available like the one where I work. It's a booklet that the businessperson can subscribe to that lists deed transfers as well as lawsuits, who is filing for bankruptcy, and so on. It's a very handy item to have around.

These two lead sources can be pursued by your office help. Let someone be assigned the job every day to check the local papers and business publications, and mark the leads. The important thing is for the checker to cull out the useless ones, leaving the salesperson the job of boiling down the original list. Every now and then leads will come from news items, and these should be brought up also. You'll go for days with dead ends and then bingo! This source of leads really doesn't take too much of your time.

RECORDS

The world's worst record keepers are salespeople. Most hate paperwork, so they have a tendency to do as little as possible and operate out of their shirt pocket. I'm no different.

It's quite important to have and keep up-to-date records to make sure that no lead or prospect is overlooked. The very fact that long time periods are an integral part of selling construction projects makes good records mandatory so that the prospect can be contacted on a regular basis according to his particular situation. It's equally important for the owner of the construction company who has a salesperson working for him to know what is going on. Employees leave and there must be some record so the job can be carried on.

A word of caution concerning your computer. Do have some security control over your database. Prospect information should be somewhat privileged.

It does no good to keep records if you don't use them. I'm into mine every day making notes and making sure nothing is being overlooked. Failure to follow up tells the prospect you're not interested in his business, and when he gets this idea, you won't have it.

WHO REQUIRES CAREFUL CONSIDERATION

This subject addresses committees, which is an area that can give the design-build contractor some very special problems. This area has to be approached and handled specially since you are dealing with a group of people instead of an individual. The person in business has a pretty good understanding of what's going on and the salesperson can respond easily. But everything changes when you have a group of people, most of whom do not have the foggiest notion of what is going on in the construction world.

People with the best intentions in the world (since they will be nice people trying to do as good a job as possible) will bankrupt you if you let them. I'm talking about church, social, civic, and even theatrical groups. I firmly believe the first time in history that three or more people met together, the first topic on their list was a building to house their enterprise. It still has to be the main subject of committees to this very day.

The first thing a committee that is interested in a building wants is a cost number. So you attend a meeting to determine the requirements in order to work up a budget.

There are some "tricks of the trade" I would like to pass on when sitting in with a committee. The first is take notice of who is the leader (and it does not have to be the chairperson), and then make sure this person knows you are a construction professional; you in essence will be selling to this person, and then let this person sell the group. At the same time, do not ignore the other members, and the best way to do this is to listen to them when they are making a point. Doing a lot of listening and making notes will make you look good.

One meeting should give you enough information to work up a budget, and I suggest this be a fast budget simply done in the office with information you have readily available. It's now your call if you present the budget at another meeting or let the committee look it over without you. Be on guard to watch your time.

Trying to sell design-build construction to a committee is very time-consuming and frankly the contract potential is not very good. Many times someone on the committee will think the best way to go is the bid route, and all your work has been for nothing.

I recommend you give a committee lead very careful consideration, and look out for number one.

CONTRACT POTENTIAL

Up to now this chapter has covered where to go to find leads. But that lead is nothing more than a name until you talk with Mr. X and dig out the necessary information to determine just what it is you have to work with.

The good construction salesperson would make just as good an investigative reporter, police detective, or intelligence agent. Why? Because people in these professions have to be able to obtain information vital to performing their job. Information is the job.

As a construction salesperson calling on a lead, you are going to have to ask questions. Never hesitate to ask the particular question that you feel will help you. The very worst the lead can do is say no, and you're no worse off than you were before you asked. You have nothing to lose.

I'm talking about legitimate questions pertaining to the project; don't be guilty of prying. As long as you're on the subject, ask away. An important fact to keep in mind when talking with a lead is that Mr. X may be able to pass on another lead whether or not he is a bona fide lead himself. Train yourself to bring the conversation around to your asking if Mr. X knows of anybody who may be in the market for a new building. You'll be surprised with the answers, and this type of lead from a lead will usually give you one that probably no one knows anything about yet, which puts you in the picture right up front.

Good information-gathering ability is not something you are born with. It has to be learned and cultivated. When Mr. X tells you he doesn't own his building, you ask if the lease is short- or long-term. If the answer is short-term, ask how short, and explain that building a new facility is not a fast undertaking.

It takes a lot of time; since he has, let's say, eight months to go, why not let you work up some quick budgets so he (Mr. X) can start getting an idea of cost? If the answer is long-term, say three years down the road, ask if he has any plans for a branch operation. Don't just accept his answer about leasing and walk out. Dig to learn something that may help you now or in the future.

In the next chapter I'll go into details of what to ask, but for now it's important to realize that the construction salesperson must be able to ferret out bits and pieces from the lead. It's the only way to determine if the lead can become a prospect, and prospects are what you are after because they are the ones who buy.

Let's quickly sum up this chapter, which is strictly about finding leads to investigate. The total number of leads that you'll be working with at any given time will be made up of leads supplied from the various sources I have discussed. No one area will provide you with enough leads that you can afford to ignore the remainder. Temporarily, under certain conditions, you may get more than is normal from one particular class of lead, but don't let this lull you into a sense of false security. The picture can change, and you'll start out one day with no one to call on.

Keep all the sources constantly in mind and pursue them. Remember you'll be playing a percentage game and it's vitally important to remain completely objective about the source of your leads. If there are things you personally don't care to do, then do them anyway. The reason you're doing this at all is for money, so don't be foolish enough to pass up leads that can provide the money.

Cold calls seem to be most construction salespeople's Waterloo. For that matter, cold calls in any field of selling seem to be the one least-liked part. I've even been told by some salespeople that they feel cold calls are somehow demeaning. To me, this is rubbish. What's the difference between a cold call and a clerk in a retail store coming up and asking, "May I help you?" When the construction salesperson makes a cold call, he's asking the same question—maybe not in the same words, but nevertheless the meaning is the same.

The reason many salespeople don't like cold calls is that they are afraid of them—afraid of looking stupid, afraid of a situation that they can't handle, afraid of being rejected, and so on. Cold calling will teach you to think on your feet, to handle situations, to meet complete strangers in all walks of life, and talk to them as well. In short, it will give the construction salesperson confidence, without which he couldn't give a job away. And after you become comfortable with cold calling, don't stop, even when you have plenty of work. It's a sure way to keep yourself fine-tuned to people. By the way, if you pick up a lead, then great; nothing like a good lead to perk up one's day, especially when you find that your closest competition knows nothing about it. So wade right in, and worry more about making money than getting your hands dirty.

Lead gathering is plain hard work; please start out with that in mind. Ask questions and always be on the lookout for a lead from a lead. Never leave a Mr. X when you are through with your initial meeting without asking if he knows of anyone else who might need your services. It's truly surprising what people know but are not aware of until someone jogs them.

In closing this chapter I've listed a chart of the lead gathering areas along with their contract signing potential for the prospect. I know it seems like a lot to remember, but after you have some experience under your belt, the chart will be imprinted in your head with your own variations added.

Lead Source	Contract Potential
1. Personal contacts	Good because lead has expressed interest.
2. Referrals	Very good because lead has expressed interest and taken the trouble to investigate.
3. Cold calls	Fair because you know very little about the call at first. Will take many calls to produce good prospect.
4. Trends in specific business areas	Fair to poor because you know very little about the call at first. Will take many calls to produce good prospect.
5. Redevelopment projects	Excellent because the company has to relocate.
6. Relocation projects	Excellent because the company has to relocate.
7. Industrial commissions	Fair because lead can be checking out many locations.
8. Rejected bids	Excellent because lead has every-

	thing together and is ready to move, but cannot for some reason, usually money. (*Note:* Move fast even at the cost of another lead if necessary.)
9. Advertising	Good because lead has expressed interest.
10. Disasters	Excellent because company will probably need to rebuild to continue business.
11. Bonus to employees; finder's fees	Fair because the person supplying the lead most of the time really doesn't know very much about the lead.
12. Zoning changes and deed transfers	Good because the lead is putting out time and money to do something.
13. Committees	Low contract potential. Be careful.

The lead sources are listed in the chart in decreasing number of leads generated; that is, you'll have more personal contacts than referrals or more relocation projects than rejected bids. This rating system is very loose, but I want to give you some kind of handle for judging your leads, so that when you have to cull out and put priorities on them, you'll have some idea of just who's hot and who's lukewarm. In other words, don't break your neck over a lead from the industrial commission when you have one from a redevelopment project and there is only time for one.

Please remember that leads are the lifeblood of marketing construction. They are where the prospect comes from; and he's the guy who hands you the check. So be on the lookout at all times for leads.

CHAPTER 5

QUALIFYING THE PROSPECTIVE OWNER

The previous chapter covered many methods that can be used to obtain leads. The problem with a lead is that it's nothing but a name on a list until you take the time to check it out and qualify it to see if the lead can become a bona fide prospect.

Not every lead will become a prospect. Again it's percentages; a lot of leads will boil down to a few prospects. I have absolutely no idea what sort of ratio you're supposed to get when comparing number of leads to number of prospects. There are way too many variables, and I feel a statistician would be hard-put to arrive at a number. This much I do know: a lot of leads will give you a few good prospects. Please note the word *good*. A lot of leads can give you a fair number of prospects, but we're not interested in just any prospect—we are after the good ones. These good ones are the leads that you prove to yourself are going to make a move; they plan to build. Granted there may be problems that must be overcome before the contract is signed; nevertheless, they plan to have some type of construction done. That's my definition of a good prospect. All the others are nothing but people for you to very nicely contribute your money and time. Be prepared to work for your prospect list. Every now and then you'll back into a good one without much work, if any; but that's going to be the exception, not the general rule.

When qualifying a lead, use the direct approach. Come right out and ask Mr. X what his plans are. Does he intend to do something? You will receive one of two answers, no or yes. If the answer is no, then you know right away this particular lead is a dead end, but never leave without asking Mr. X is he knows of any other likely people whom you might call on. Remember, never pass up a chance to fish for a lead.

The no answer is cut and dried; you know immediately where you stand. When the answer is yes, then you think you have a real honest-to-goodness prospect to work with. Well, I hate to burst the bubble, but this isn't necessarily so.

Our Mr. X who's just answered your question with an emphatic *yes* could be telling you yes while in truth he doesn't have a clue about what's involved. He may have a burning desire to build that new facility, and just because he wants to, he thinks he can do it. It's up to the construction salesperson to dig out enough information from the lead to determine if he is a good prospect.

You won't know when the lead answers yes if the yes is good or bad. The best method to use is to assume it's a bad yes and start right off digging for the needed details. If the yes is a good one—that is, Mr. X has the preliminaries put together, and it's quite obvious to the salesperson that this lead knows what he's about—then you lose nothing at all at the beginning by assuming he's a bad yes. You will quickly pass through the preliminary qualifying and get down to the nitty-gritty. On the other hand, most of the leads will take qualifying starting at square one, so we'll start at the same place.

SITE

The first requirement for a contract is for the lead to have a site. Absolutely nothing will happen unless our Mr. X has a piece of land on which to put the building—be it a new building or simply an add-on to an existing building. Of course, the site is not a requirement for interior work or repairs, but most of the time it will be an important part of the process. This is as important to selling construction as simple basic addition and subtraction are to the science of mathematics. I stress this point so strongly because I've noticed that construction salespeople have a tendency to become so wrapped up with a project that this simple, basic rule is overlooked.

While you're talking with Mr. X, you steer the conversation around to the site, and Mr. X tells you he doesn't have one yet; as a matter of fact, he hasn't really started looking around, but he's going to do this and that when he locates the right spot, and so on. When you hear words to this effect, realize Mr. X is not a good prospect and you are not going to devote very much time to him, if any. I'm not saying drop this lead; I'm saying back off until the lead is serious, and lining up a site is the first step in this direction. Keep the lead on your list and check back every few months to express interest and keep your name in front of Mr. X. After all, the lead can suddenly become serious and buy a site, and you'll want to know about it from Mr. X, not the grapevine well after the fact.

I'm sure you're aware that nothing is simple anymore, so there are variations of this no-site rule to know about and to know how to use in your selling. What about the lead who has no site but who you feel is not daydreaming? He's going to do something. I want to be very clear on this point. I'm not saying forget a lead with no site; I'm saying don't spend any time or money except what you can easily provide with no trouble to yourself.

A lead came to me in the form of a personal contact. The lead was thinking about building a new facility, so I immediately jumped on it. It didn't take me long to determine that Mr. X was going to build; the problem was that he was undecided about just where to relocate—hence no site. I told Mr. X there was little I could do until he had a site because the prices would not be good indefinitely. He did give me the general specifications for his new store and asked for a handle on the cost, just a ballpark price. I have done quite a few jobs like the one he wanted, so it was only a fifteen-minute job to budget the cost. I didn't tell Mr. X this; I left him with the idea that it would take a little time but I was willing to do it if I could work it in during the next couple of weeks. I wanted Mr. X to think I was really putting in time to help him. I was simply selling myself. Now that's all the pricing I was going to do—I didn't care what else was asked of me.

I returned with the budget a few days later, spent several minutes going over it, and then left, telling Mr. X when he found a site to let me know and I'd be glad to look at it to evaluate hidden construction costs. This would keep me in the picture and would begin to make Mr. X dependent on me.

Six months went by; then Mr. X called me to check out a site with him. I might add that during the six-month period I had called him several times to touch base so that I could stay in the picture. Well, we checked out the site, and it had nothing but problems—sewer, setback requirements, plus a road exposure that was terrible for a retail business. Site costs would run up the overall costs, and maybe he should look around again, I told him. This went on for a year, and all I had invested was a little time. I felt under the circumstances that the small amount of time was worth it. At no time did I do any more estimating. I was positive Mr. X would build when he found the correct site. Also, he was calling me, which showed interest. Well, after a year he finally located a site that filled the bill, from his business viewpoint and with respect to the construction cost. The real estate agent was a friend, so I asked him to keep me informed as to progress made on buying the lot.

Three months later the agent called and told me Mr. X had just closed on the site. This brings up a point that's a complete mystery to me. Why does it take so much time to settle something that's on the market to be sold? Land deals seem to involve no concept of time. Anyway, with the site in hand, Mr. X became a prime prospect, although granted I had somewhat of the inside track by virtue of my prior work. To make a long story short, it turned out to be a nice job.

My point is that Mr. X didn't have his site but did meet all the other requirements, so I continued to follow up and work with him as long as the costs to me were rock-bottom minimum. I wasn't going to provide plans, price letters to banks, and all the other help necessary to get a project off the ground.

I mentioned earlier bringing the interview conversation around to the site. Just how the salesperson goes about asking for information is important. Think back to the first chapter and low profile/soft sell. Now with this in mind, imagine how you would look to Mr. X if, when you asked him if he had a site and were told no, you then closed your briefcase and turned to go, saying over your shoulder, "Call me when you have one." Qualify a lead like that, and you can bet he'll cross you off his list.

The way it should be done is to build the correct question right into the flow of conversation. For example: "Mr. X, in order to work up a meaningful price, I'll need some information on your site." When you give the reason for needing the facts. Please remember Mr. X will not be familiar with all the little details but will start to realize there's more to building a new facility than just saying, "Let's do it." You are selling Mr. X on your competency. So you go on, "It's important for me to know where the sewer is located so I can check with the city to see if there will be any problems tapping in."

The reasons given to Mr. X for asking for site information can be numerous; just use something. Surprisingly enough, I've found that most of the time there really are legitimate questions that need to be answered.

At this point Mr. X tells you he has not located a building lot. Don't let on for one second that warning bells are ringing inside your head; calmly go on with the meeting. Move ahead to the other qualifying questions. Make no hasty decision yet concerning this prospect. You need all the facts first, so go ahead with your diggings.

After you're sure nothing will happen until Mr. X locates a site, you have to pull back without giving the impression to Mr. X that that's what you're doing. The tack I always take in this situation is to tell the lead that until a location is picked out, it's very difficult to work up any costs that will be worthwhile. I tell him construction prices are always in a state of change. It will do no good to supply building costs without site costs because the bank is interested only in the total cost. Then I throw in the hooker: I tell him I would be doing him a disservice to give out what will become worthless and, worse yet, dangerous information.

At this point I offer to give him a handle on the building cost if he will use it only for personal information and take note of the fact that it will be on the high side. Now I offer this mainly to keep myself in the running if, in my judgment, this lead has promise. This judgment call is just what the word says; it's strictly based on my feeling about Mr. X, how he comes across, what has been done to date—in short, the sum total of all the small parts that come out in a qualifying interview. Here the construction salesperson is completely on his own. He makes a decision and then proceeds from there, often hoping he's done the right thing. This is quite hard to get out of a book. I can try to give you some guidelines, but that's all; the rest is up to you. And if you're working at it, you'll make mistakes. Don't let them worry you. Mistakes come from doing something; mistakes aren't made if you're not doing anything.

Sometimes "no site" can be used to help sell yourself to Mr. X. I think many times our lead hasn't done anything about his site because he really doesn't know how to go about it. He has some ideas, but he may hesitate to become involved with a real estate broker. Many times he's so wrapped up in his business that he won't take the time to drive around and see what's available. Here the construction salesperson can be of some help. Tell Mr. X some of the details concerning water, sewer, gas for heating, and so on, and then recommend a good commercial broker to work with him. If he says okay, he'd like to talk with him, offer to call the broker yourself; tell Mr. X that you'll be able to give the broker construction facts that will help the real estate person do the homework so as not to waste Mr. X's time with unsuitable locations.

This strategy does several things for the salesperson. It lets him stay in the picture and also enables him to call the broker with whom he wants to work. You give him all the information concerning Mr. X, if possible even going so far as to give the broker your readout on how Mr. X likes to be approached. The really important fact is that you have called and given the broker a good lead. Since all of us travel a two-way business street, the broker will want to repay the favor in the future.

One more excellent point to bear in mind about contacting the broker will be the possibility of developing an ongoing relationship that will keep you informed about the progress being made on finding the site. He might even learn if another contractor comes into the picture and pass that information on to you. Forewarned is forearmed. Be sure to ask the broker to contact Mr. X right away and let him know that he (the broker) is starting to work on his site. Then several days later check with Mr. X to find out if the broker has indeed made contact with him. There are two reasons for checking back: one, to keep the ball rolling, and two, to show Mr. X that you're a hustler, which is just another small step in selling yourself. Never let up on selling yourself; it has to become as natural as breathing.

Without the site there can be no contract. If the lead doesn't have his site, then be careful about spending time and money on him. Do offer to help by bringing in a broker (whom you can work with); then step out of the way, not out of the picture; and keep tabs on the situation, because when Mr. X obtains his site, it's time to go to work.

FINANCES

The second most important fact in qualifying a lead is money. I've always heard that love makes the world go around. I don't buy that—it's money!

Without money our lead is a dead end. It will take money as well as a site to make this lead into a prospect who can sign a contract. There is one major difference between the two. The site location can be critical—not any old place will do—whereas any old money will do nicely.

Bringing up the subject of money in the conversation is a little tougher than bringing up the site. The best approach to opening the door on the money question is to assume that Mr. X has no problem at all. He will need some information to satisfy the lending people, so you offer to help with what's needed. This will usually start the talk about finances, and from there it's an easy matter for the salesperson to ask some direct questions under the guise of gathering information to help. You are doing just that, but at the same time your prime concern is whether Mr. X has his ducks in a row so he can pay the bills.

Nine times out of ten the lead will have talked with his banker and will have some idea of what he can expect. Usually a good businessperson will talk to his banker first, before proceeding any further. Without plans and contract costs to supply to the leading people, everything is verbal and tentative. Nevertheless, the lead will have some idea about the money. My experience has been that Mr. X won't hesitate to tell the construction salesperson about already talking to his banker and that there is no problem. I've also found that's just about all Mr. X will say about his money; he has it lined up, and that's all he'll tell you until he knows more about you personally.

The gray area is what you have to be prepared for now. It's sort of like the twilight zone where it's very hard to prove something that's important to you. Mr. X tells you he wants a building and you ask for details, which are forthcoming. Mr. X tells you he has a site, and you ask where; the answer is a simple address—you are dealing with something tangible. But, how do you prove to yourself that Mr. X has the money arranged as he says? Start prying too deeply into his finances, and you'll probably push his *off* button (remember that from Chapter 2?), as well as make Mr. X somewhat upset. No one likes to be doubted with respect to money, so you'll just have to accept his word.

The only defense you have is to try to find some circumstantial evidence that could give you some idea of whether Mr. X indeed has the money. The best way is to offer to supply some needed information that the lenders will require. Sometimes Mr. X will start discussing some of the financial aspects. Hopefully you'll pick up enough to put your mind at ease. In any case, you'll be in a gray area until the project moves to the plans step. The good construction salesperson learns to live with this problem; it comes with the territory. It certainly won't hurt if you are able to make some discrete inquiries as to how Mr. X stands financially within the local community.

Mr. X has the money, or at least you have been told he has. But what about the lead who has a site and no money? You'll come across this situation every so often. I'm talking about a lead who wants and needs a new building, has optioned a site, but for some reason can't get the money to make the deal fly. The lead will probably not hesitate to tell you his tale of woe, for he's desperately casting about for help with his problem. This type of lead can be a good prime prospect under the right conditions.

The construction salesperson will have to know something not only about design-build but leasing the new facility back to Mr. X. I'll cover the details of how this is done in the next chapter. For now, we are only interested in qualifying the lead.

You may be asking, why not assume that the lead has the money to build and not use this as a qualifier? Well you can, and a great many times it will be perfectly all right. But you have no defense against the big talker who likes to think of himself as a high roller. I use a handy rule that is accurate almost every time: beware of the easy sale and the lead who casually brushes aside talking about project finances.

The money is why you're out there talking with this lead, so find out all you can. But at the same time remember this is probably the touchiest subject you'll discuss with Mr. X.

REASON FOR BUILDING

Our person has a site and his money, so let's look at another important area. Why does he want to build? Think back to Chapter 2 and why businesspeople build new facilities. They are forced to. They may be forced by being in the path of redevelopment and relocation projects, or they may be forced to expand because their competition is expanding and they are afraid of losing business. It could be that their lease is up and they can't renew it. They may have no choice but to build the new warehouse to be able to handle the new accounts they've just taken on. Regardless of the amount and type of force being applied to the lead, Mr. X will only move when he has to.

It's important to the construction salesperson to find out

what the moving force is. It will help him qualify the lead, to see if there is a good prospect to be had, and also help the salesperson determine the time factor involved if the lead does indeed look like a good prospect.

The following example shows exactly what I'm talking about. A few years back I contracted with a local firm to build them a new office building. It started as a very general lead from a friend. The lead was that the local Air Force base was expanding the safety zone at the end of its main runway. Thus, I had learned of a relocation situation through a personal contact, so that two lead sources were actually combined. Anyway, there was only one good lead to come from this area; so I went right to work. The lead checked out okay in every qualifying category, with one small exception concerning the time for building.

There would be a time lapse until the Air Force actually closed the deal. With this information I told Mr. X there was no need for me to do anything until then. We left everything as it was at that time with the clear understanding I would stay in contact to follow up. It took almost a year for all the final details to be worked out, at which time I went to work to put the job together.

It's vitally important to know the details of why Mr. X has to build. It helps the salesman judge exactly how to proceed with the project, and at the same time the salesman can determine just how serious a lead is. As in any other selling business you get the lookers, the people who may be thinking it would be nice to have a new building and who want to know what it would cost on today's construction market, which leads us into the next topic.

INTEREST

There is a way to tell if a lead is really serious or out window-shopping. There's not too much to it: you give the lead some small duty to perform. He's serious if he does it, and not serious or at least questionable if he doesn't do what is asked.

The salesperson should be able to get a feel for the lead as far as interest is concerned; this will come with some experience. It helps if you can have a readout before you have to "assign" a duty to the lead. The problem is that this requires a return call or maybe several until you can make your determination.

Okay, you say, what do I do? Simple. Ask the lead to get you some information on the site from the real estate broker, or the surveyor if one is involved. Ask him to check on a zoning question, or perhaps a decision about some detail of the building that you need to know before the price can be estimated. Here's a sure method that I've found most reliable: I ask Mr. X to give me an hour's time so I might show him some buildings for him to get a better idea of what we're talking about. If this is really impossible, and it will be with a retail businessperson once in a while, give him directions and suggest that he drive by on a Sunday afternoon and have a look. If he never does, then forget him.

Let me give you a good example. I was trying to qualify a lead who thought he wanted to build a new retail store. Everything checked out: he had a site; the money was there. But Mr. X didn't seem to want this new store quite badly enough; something didn't ring true, so I "assigned" him a duty. I had been involved with two stores in his type of business earlier. I explained in detail how these two buildings were different from each other and were still used for the same purpose, and that I wanted him to see them. One seemed to be exactly what he had in mind. I gave him directions, and it was only a 45-minute drive from his home, so he planned to take a look on the following Sunday. The next Tuesday I stopped off to see what he thought about the store. Well, he hadn't gone over to look at it.

If you were going to build something, and you found out where you could see what you had in mind, would you go look at it if you were really interested? You bet you would! Mr. X wasn't interested enough to go look; so he must not have been really interested at all. I dropped the lead like a hot potato.

The interest area is also a gray zone. It's quite hard sometimes to work up a readout on the lead; and, believe me, nothing is guaranteed in construction selling. I've found from experience that you will seldom disqualify a lead from lack of interest. I think every businessperson would like to talk about a new store or expanding the warehouse. But it's important not to take interest for granted; when you do, your chances of being used increase. This is definitely not good for the construction salesperson.

I follow a general rule on this subject and think about it enough to keep my perspective: do all you can to maximize the number of good prospects, and do all you can to minimize the manipulators. You will be manipulated! It really will happen, and it will probably upset you, but don't let this unpleasant side of construction selling turn you off. It comes with the territory because you're dealing with individuals before they are completely ready to build. The whole idea of qualifying the lead is to cut down on the instances of just being used. Remember, there is nothing at all that says a good prime prospect will not nail you to the wall.

When I first started selling construction, I was Mr. Eager Beaver and gave the lead/prospect everything he asked for, most of the time with no questions asked. My "being used" percentage was pretty high. As I gained more experience, this percentage steadily dropped. Another factor has had a small part in this percentage drop: age. I honestly can't say much for becoming older except that it seems to help your business image.

There are two ways of being used. The first one is almost impossible to avoid. The lead checks out and you start selling; you take all the information, do your homework, and put together a price. The prospect takes one look at the numbers, has a fit, and discards the entire project. You know there's no way to cut costs and meet the requirements and the prospect won't compromise; so there you are. All that work for nothing, but—and this is an important but—neither one of you knew this in advance, so I suggest you make it a point in such cases to part friends because your defunct prospect may decide to make some changes later. Better yet, the construction salesperson should keep in touch. A phone call every so often will do the job.

In the other type of manipulation the prospect is using you to obtain every scrap of information, drawings, and whatever else he can get. He knows all the while that he doesn't plan to build now, but he doesn't hesitate to take advantage of you. Sometimes a person will do this and see absolutely nothing wrong with it. After all, you did call on him and ask for the business; and since you offered, he intends to get all he can. This type of person, I've found out, will let you run your legs off but will do nothing himself. Assigning him a duty to perform will flush him out every time. It will not get done and the excuses you'll hear can be somewhat entertaining.

Sometimes the interest factor can be combined with other areas of qualification. For example, the interest emphasis may be put on the site so that you're checking two qualifiers at the same time, which will give the salesperson a good handle on the lead. There are other ways to combine qualifiers, all with the interest factor. Again, for example, if I get the feeling the lead is a blowhard, and he tells me money is no problem, I respond immediately with how glad I am to hear that; and I say that since he's a good businessperson, I know he will want any general contractor bonded who does work for him. That being the case, I'd appreciate a copy of his loan commitment for my bonding company so that there will be no doubt about the money back at the home office of the bond company. If the lead is really using you, he probably hasn't even talked to his banker, much less obtained a loan take-out letter. Mr. X at this point starts backing away. His hand has been called with some heavy stuff. He won't like hearing about home office bonding people being interested in him. He will realize he's dealing with a professional, not some rinky dink he thinks he can mentally manhandle. I've used this interest "dig" quite a few times and it works every time. I've been surprised also: twice the lead handed over a copy of his take-out.

I'm sure with time the new construction salesperson will be able to feel comfortable in this area and have confidence in his decisions. You can't be right all the time, but when the percentages are totaled, you'll be ahead of the game. Remember, you're talking about business with Mr. X, so don't be afraid to ask questions about his business and the planned facility. You can never have too much information.

I know I've painted a grim picture, and it seems you'll be out there on the streets looking over your shoulder all the time. Not so! Everything has a price, and this is the price you pay in selling construction; but look at the other side of the coin. There will be many profitable projects done with first-rate individuals; some will even end up as friends. I sold three 18,000-square-foot roller-skating rinks to one owner. This owner is now a good friend, and every year my daughter, while in high school, received a free pass for one year at any of his rinks. (Of course, my friend is also a good businessperson, since every time my daughter went skating, she managed to fill the car with skaters who had to pay!) So don't be downhearted; there are plenty of good times in selling construction.

To sum up the interest factor, if Mr. X doesn't display any, then move on to the next lead; without interest, there's no sale.

TALKING TO THE DECISION MAKERS

More time has been thrown away talking to and working on the wrong person than in all the rest of the qualifier problems put together. The name of the game is selling percentages, but here the construction salesperson has a mental block, especially the new one.

I can understand perfectly how you can get caught up and carried along with the momentum of putting a project together. You make a call. Mr. X is easy to talk with, and he's planning on a new warehouse. You ask about the site, and he takes you out back and shows you the building location. You ask about the money, at the same time knowing the firm enjoys an excellent credit rating. You establish the reason for building—taking on two new lines of equipment. There's interest shown. When all these subjects click, the good construction salesperson is like a racehorse in the starting gate. And when the prospect asks for a proposal as soon as possible because he needs the new building by such and such a date, then look out—our salesperson is cranking up! It's exciting to put a beautiful deal together and very easy to become intoxicated with it. So our eager salesperson puts together a proposal and presents it to Mr. X. Mr. X looks the information over and casually says, "Looks okay. The price is within our budget also." Our salesperson is ready to whip out his pen so Mr. X can sign. Then Mr. X drops a bomb on him: "I'll get this right off to Mr. So and So; he's the man in our main office who'll make the decision."

At this point, the salesperson has just said goodbye to a whale of a lot of work, time, and money. This can happen to you if you're not careful. I know from painful past experience.

There will be exceptions. For example, the branch manager is a friend and he asks you for a proposal. You don't mind working one up because you feel you'll have a friend to help you. These exceptions will be few and far between. As for the rest of the leads, keep both eyes open, and know just what it is you're getting into. Beware of branch managers; they are the best manipulators the construction salespeople can encounter. I mean nothing personal; I have some close friends who are branch managers of out-of-town firms. It's in their nature to use the construction salesperson because he can help them. They all want more space, office, storage, whatever, and they persistently hound the home office. When the salesperson walks in and offers to give the branch manager some more ammunition to send to the home office, the manager takes him up on the spot. And I would do the same thing in his place, I'm sure.

The best defense is to find out right away who it is you are talking to. Use a little finesse. Don't push the *off* button. I've found that if I ask someome who obviously is not in charge who is, and then the owner or manager is pointed out, he or she helps me through what could become an awkward situation. Anyway, be certain to whom you're offering your services.

When you determine that Mr. X is indeed the manager, go right on with the qualifying and requirements as if the lead is a prospect. Tell Mr. X you'll be glad to work up a proposal; then ask him when the person from the home office will be in town—it would help to know this so that you could schedule your time. Also be quick to add that if time is very important, you'll be glad to contact the home office. You're telling Mr. X in a nice way that you're not going to do business with him. You expect to deal with *the* person.

I have never sold a job where I couldn't talk with the decision maker. I have sold several very clean, fast-to build, and profitable jobs to out-of-town companies where I could make my sales presentation to the person who could make the decisions. I'm speaking from experience; I was taken to the cleaners so many times that even my wife told me to wise up. Now I don't make proposals to managers unless it's immediately clear that I talk to the decision maker.

You may feel that you're walking away from some business, but you're really not. If you are lucky and get such a job, there's never enough money in it to begin to pay you back for all the work you put into the ones you missed. Percentages again—that's all selling is; and the successful salesperson knows how to play them to his advantage.

I know you'll uncover a lead who turns out to be a supergood prospect, with the only problem being you're dealing with a manager. It looks good, and you feel you should give it a shot. Go ahead, that's fine—except don't let it take away anything from a better prospect, even if the prospect doesn't offer the same inducement as the manager. Look at the situation as investing some money in a high-risk operation. Invest only what you can afford to lose. Play the percentages; better yet, believe in them. A good friend of mine, who's an air force pilot, told me the hardest thing he had to learn about flying was to look at his instruments and *believe them*, no matter what he thought personally. The same goes for the construction salesperson. Learn to read the signals and then believe them.

There's yet another pitfall to watch out for: the father-and-son team. It's hard to determine who to talk to. If you start out with the father, chances are you'll be okay; but if the son is the one wanting all the information, be careful. He has to be considered in the owner category, but chances are he won't be the one who will make the decisions. He'll also be eager to do something—he'll be interested in making the business grow, so he'll jump right on the services you can provide. Then the father will say no for some reason or other. Be prepared also to say everything twice with the father-and-son lead; for some strange reason you can never see them both together—one always has to watch the store. Just try to find out all you can about the father-and-son lead. It's important to talk to the decision maker without making the other one (usually the son) upset because he's not the one running the show. You've got to be an excellent diplomat in this situation.

To sum up, if you cannot talk to the person who will say yes or no and sign the check, you're wasting your time. This qualifier is just as important as any of the rest I've explained, maybe even more so; for it is the one that's the easiest to overlook and/or want to overlook, so that the salesperson gets sucked in.

PERSONAL PROBLEMS THE LEAD MIGHT HAVE

This subject is not a qualifier in the same sense as the preceding ones, but knowing about the lead's personal problems will help the construction salesperson make better use of his time. Note that I'm not talking about personal problems in the sense that immediately comes to mind; I'm talking about the personal problem that is talked about in public.

Here's a good example. I worked with a good prospect who planned to build a 20,000-square-foot metal building, 34 feet high, to be used for boat storage. In the course of working out the requirements, Mr. X mentioned that he wouldn't need the information for three or four weeks. Mr. X was due to have an operation and would be out for almost a month. With that information I'm able to work my schedule around the proposal to my advantage.

Most of the personal problems that I've had to contend with have involved sickness, and I'm sure your experience will be the same. Sometimes the construction salesperson picks up a piece of information that really can be a semiqualifier. Always ask the lead what his time schedule is; usually here is where the problems will come to light. At the same time, if you hear something about Mr. X having trouble trying to buy out his partner, and that he'll not be free to pursue the project until that's taken care of, which should be in such and such a time frame, become a little cautious. Don't jump up and run, but become wary.

Remember to be tuned in to anything that will help you. Pay attention when Mr. X says something about personal problems. You may even be subjected to hearing about his upcoming divorce or whatever. If so, listen; you can never have enough information.

A lead is nothing more than a name until you qualify that lead into a good prospect. A good prospect is one who passes all the qualifiers and plans to build. The qualifiers that you will use are site, finances, reason for building, interest, talking to the person in charge, and personal problems.

It's imperative that the construction salesperson understand how these qualifiers are used and why. Their use should become second nature, and with experience it will be possible to qualify a lead in a general way without calling on him or her personally.

When a friend passes on a lead to you, ask if Mr. X has his site and money. Your contact may have the answer and save you some time. But don't depend on him too much. It pays to contact the lead even though you know he doesn't have a site, to show him you're interested so that you can keep your name up front. Also, you may be able to recommend a solution to the problem. As a matter of fact, a solution for the problem would be an excellent reason for contacting the lead.

Asking questions from every source, digging out details, and putting all the information together is going to be second nature to the really good construction salesperson. And I'm sure you are just like me: you'll have to learn to develop these techniques. No one is born with them. Some people seem to be more people oriented than others, which helps, but it's not mandatory. Hard work is!

Learn how to qualify that lead; if you don't, all you'll be doing is spinning your wheels. Nothing else is as disheartening as working your behind off with nothing to show for it. This chapter will help you contain such wasted effort to a minimum; it will also help you improve your contract-signing percentage by allowing you to work prospects who plan to build. All you have to do is convince them to let you do the job.

Read on. The next chapter tells you how to convince them.

CHAPTER 6

SELLING THE DESIGN-BUILD METHOD

The purpose of everything the construction salesperson has done so far is to get him to this point. It's as if he has been in basic training, and the time has finally arrived for him to move up to the front lines.

From here on the order takers are in trouble, because it takes a results-getting salesperson to convince the prospect to give him the contract for his new facility. The sorriest of salespeople can do with ease what has been gone over so far. It's really fairly simple to dig up a lead and qualify it. It mostly just takes hard work. But now it takes brains as well as hard work to sell the prospect. Prospects are what you'll be working with now—no more leads.

An old sales manager I worked with when I first started out told me right after I went to work, "Never forget, boy, nothing, absolutely nothing, will happen until you sell something." I've never forgotten what the old gentleman told me; but more important, he's dead right. Nothing does happen until somebody sells something!

Remember in the first chapter when design-build capability was brought up? It's now time to put it to use, because without it the qualified prospect is nothing more than a name on your list. Up to now, with minor exceptions, you have been a one-person show. But from here on you have to be able to pull together costs, estimates, budget proposals, and final contract figures from information and drawings that have been generated by the design-build team.

SECURITY

So far the construction salesperson has not had to rely on any outside help, except for other employees within his firm. Now when he starts selling Mr. X, he will have to bring in outside help. For example, there may be a zoning problem the salesperson needs to check, so he visits city hall and talks to people about the situation. He may see a health department official if the prospect is going to build a restaurant. Subs will have to be brought in to look at preliminary plans and quote prices. At times a sub will be asked to quote on a design-build plan that requires details about the prospect. In short, you are putting the job prospect on the street. Don't underestimate your competition; if they get wind of the prospect, they will be all over him. This will make your job of selling that much more difficult, as well as cut down on your chances of signing up the job. Percentages again: the more contractors working on a prospect, the worse your chances of selling the job.

The construction salesperson should concentrate on job security and not worry about what anyone thinks about being security conscious. There are a number of ways to work the job without giving out any information about who the prospect is. Preliminary drawings are usually a dead giveaway. It's only natural to cover the sheet with the prospect's name, if for no other reason than to impress Mr. X. Don't do it; fight the impulse, and leave it off. Also, while we are on drawings (and I'm talking about preliminary drawings or sketches that you will be using to get budgets from subs, not architect-prepared plans; they come after the prospect is signed up, and then security is not a problem), don't spell out what the building will be used for. For instance, if you are working on a retail auto parts store, don't spell it out. List the job only as "retail store." That's hard to pin down; "auto parts store" can be run down by a smart construction salesperson in a matter of hours.

You may ask, how does the information leak from the sub? You have a good relationship with the sub, and he's not going to run his mouth. I agree he may not, but chances are he'll have to contract a supplier; there will be salespeople calling on him so he can put together his price for you. Your sub may be able to keep his mouth shut, but he may run a sloppy office as far as security is concerned. All an outsider has to do is look around the desk or drawing table, spot your drawing, and right then he's in the know, and your prospect is on the street. Contractors' offices, both general and sub, generate a lot of activity. There are people coming and going all the time, and anything left out becomes general knowledge very rapidly.

Security was a one-person problem when you were working alone; when you have to bring in an outsider, security is still up to you. So, as we've said, don't let your drawings give you away. They are one of the best ways of revealing information, since it is recorded on paper to be read by all. A few simple precautions will go a long way in this case; use your head and don't be careless.

Also, when talking to anyone, including subs, refer to the job in general terms. The prospect simply becomes "the owner" when I'm discussing something that pertains to the job.

General contractors are extremely lax in internal security. Internal security is very important when you're doing design-build selling. The best rule to follow when you're trying to sign up a prospect is right out of the spy handbook—what they don't know they can't tell. Although I'm sure you don't have to worry about industrial espionage, what concerns me is the occasion when the steel foreman is having a beer with his opposite number who works for your competitor. Shop talk will almost invariably occur, and all your person has to do is say that his company is working like crazy to sell a new warehouse to a moving and transfer company. It would be said in all innocence, but the point is there's an excellent chance that the word would get back to your competitor. I know this sort of thing goes on. The people in a certain trade all know each other. They seem to float around working first for this company and then that one. Often these people will work together for a while, then for competitive outfits, depending on the workload. But they still know and see each other as well as discuss their work.

You want to make sure your prospect is not the main topic of discussion at the local tavern next week. Any employee can unknowingly blow the whistle on you. The secretary going to the post office or the bookkeeper making a deposit at the bank can pass the time with someone from another construction company. I feel the real problem is that construction company employees are a little more concerned about job security than people in other jobs; therefore, the uppermost topic is, how's your workload? They're not prying, just interested. If one firm's contract load is off, they want to be reassured by hearing someone else's is also off, which probably means the industry is off a little. Misery really does like company. On the other hand, if they are loaded with work, they want to brag. The net result is that the workload is talked about often and in such conversations that cat can be let out of the bag.

It's impossible to keep your employees in the dark, but you can impress on them that their job may depend on their keeping their mouths shut concerning prospects and upcoming jobs. Maybe it wouldn't hurt to suggest they listen a little closer to other people and bring in a lead so that they might be able to pick up a nice bonus if a sale is made.

I feel the best way to handle the internal situation is to explain to employees that what they hear in the office is confidential. Next, work on a need-to-know basis. If you have an estimator to work up the prices, make sure he doesn't run through the office telling everyone. At this stage of the selling procedure, he's probably the only one who needs to know anything specific. All the rest of the office needs to know is that the estimator is working on a building price. As the selling process moves ahead, other people will be brought in. The bookkeeper will have to charge the estimator's time to something, so this person gets in the picture. Finally, the proposal has to be typed, and suddenly the typist knows almost as much about the job as you do yourself. Hopefully Mr. X signs the contract soon after this.

Along with the office help, you'll have the chief field people stopping by the office, and it's only natural for them to have a cup of coffee and chat—with you, or the estimator, or with other employees. You may stop by a job and find that the conversation turns to work in the pipeline because the foreman is vitally interested in the future workload. This will often happen if you're working as a construction salesperson for someone else. Anyway, it's hard not to let the key field people in on what you're doing. Tell them anything you like as long as you don't compromise the prospect. This takes some diplomacy because you don't want the foreman to get the idea that you're hesitant about talking to him. I can't tell you what you should do. Only you know the foreman in question well enough to handle the situation.

There's a natural tendency to want to give your people a lift by telling them what is going on when the future looks bright, but don't tell all if it might come back to haunt you. Learn to think about your external and internal security. Good prospects are not that easy to come by, so give yourself every break to sell the job. The same sales manager I spoke about a few pages earlier told me something else that I'd like to pass along. He said, "You never learn a thing talking." And I'd like to add my own rule, "You'll never give anything away when you're listening."

For your own sake, prospect security is extremely important, so practice it.

PROSPECT LIST

Before you start out the door to sell a contract, it will help you a great deal to know how you're planning to use your time with prospects. I've found the most efficient method is to keep a simple list of the prospects in order of their importance. A glance at the list always helps me keep my perspective.

Your list will always be in a state of change, depending on the progress, and in some cases, lack of progress of the negotiations. Remember, your list should be made up of people who are planning to build, so that everyone who is on the list is a possible future job; and you should divide the list into categories, depending on just how good a prospect is. For example, I use three categories:

1. *Fair*—This prospect may have the desire, but money problems, zoning difficulties, or some other roadblock could be in the way. Nevertheless, the prospect may resolve the problem, or you may help in some way to move the project along. This prospect is only a fair one.
2. *Good*—This prospect has the desire, no problem with money, and even a site (in some cases there will be more than one site to contend with), but for some reason he is not completely committed to moving ahead. The state of the economy is usually the reason, or there could be personal reasons.
3. *Excellent*—This person has his act together and has made up his mind to build! He may be the prospect who is forced to move. No matter what the reason, the job will be built.

One problem the construction salesperson has to guard against is becoming so wrapped up with a prospect that other details that should be taken care of drop out of sight. He must constantly be checking the overall selling picture. He has to be a juggler of sorts, to keep several prospects in various stages of work as well as following up with leads, and all the other nitty-gritty details for which he is responsible.

Work from a prospect list. Keep it up-to-date and use it to play the percentages. Always go after the prospects with the best possibilities for success. There will be times when the list will be short and you can work with all of them. There will also be good times, and you'll be able to pick and choose.

SELLING YOURSELF

The very first thing the construction salesperson has to do is sell himself to Mr. X. The prospect has to have confidence in you and like doing business with you. He has to believe the salesperson is going to look after his interests and protect him when necessary. You will become *his* contractor; Mr. X will call you "my" contractor just as he speaks of "my" attorney or "my" CPA.

Let me point out that much of the first-impression part of selling yourself will be directed toward a lead because the two of you will be meeting for the first time. The remainder of selling yourself will be directed to a prospect; it involves your personal performance while putting together the project and trying to sell it.

First Impressions

From the very first time you meet Mr. X you start selling yourself, and you never stop. There are as many ways to sell yourself as there are people. The bywords are *low profile, soft sell*. Everything you do is within the framework of these words.

The very first real contact Mr. X has with you is in the impression he forms when he sees you. First impressions are extremely important; therefore, I let Mr. X see exactly what he expects to see, a man who works in the construction industry. Let me ask you a couple of questions. How would you like to be sitting in the examining room waiting for the doctor when the door opens and, instead of a person wearing neat business clothing under a white coat with a stethoscope hanging from a pocket, in steps someone needing a shave and dressed in jeans, t-shirt, and sandals? How would you like to find your banker dressed in casual clothes for a business appointment? What would your reaction be to these two situations? I imagine the same as mine: you'd start thinking about changing doctors and bankers. Your confidence would be shaken. These two individuals might be competent beyond belief, but I would hesitate to give them a chance, since they didn't look the part.

The point I want to make is that you should look like a construction person who is in management. I've developed some ways to dress that I feel give Mr. X just what he expects. I wear clothes that are practical outdoors yet perfectly acceptable in a place of business. In summer my standard uniform is neat khaki pants with a short-sleeve dress shirt. Depending on whom I plan to see, I may wear a tie and coat. There are many sport-type shoes that are nice-looking yet practical since you will at times be on a construction site. The khaki pants and sport shoes give the construction touch, while the dress shirt, with or without tie and coat, give a management took—a perfect blending, I feel.

For cold weather, whenever possible, I wear khaki pants with a dress shirt and corduroy coat. Again the tie is optional. When the weather is bitter cold, I wear slacks, dress shirt, and a sweater plus a car coat. I know there are dozens of ways to accomplish our purpose: to give Mr. X the impression he is looking at a "take-charge fellow" who works in the construction industry. And I'm certainly not saying that everyone should run right out and buy a new construction wardrobe. I am convinced that appearances are important, and I personally prefer to wear clothes that I feel enhance my business image.

Of course, when the occasion calls for a suit, I wear one; and many times it is simply good taste to do so. My way may not be your way of operating. The super-big construction firm will have many executives, and they'll probably look like executives. My way seems to work well for the small-to-medium construction company where management is included in all aspects of the jobs. If you plan to have salespeople working for you, you should make clothing suggestions.

I almost didn't include this section on appearances, for fear of seeming somewhat presumptuous—the very idea of my telling you how you should look to the prospect! But I feel very strongly that appearances are important, and the subject should be discussed in detail.

Several of the subjects below were discussed in Chapter 2. I'm bringing them up again in order to have everything in this area "under one roof," so to speak.

The Handshake

Give one that is firm, brief, and friendly, I always like to lead; I put my hand out first so that there's no awkward hesitation while each party waits for the other to offer his. A psychologist probably could talk for hours about this. Many owners, especially in smaller shops where they work with their hands or supervise people who do, seem uncomfortable when shaking hands. Maybe they have trouble relating to total strangers. Be alert for the feeling, and try to put Mr. X at ease as quickly as possible. Remember, regardless of the condition of Mr. X's hand, if he puts his out, shake it!

Addressing Mr. X

I have a hard and fast rule about addressing Mr. X. He is always Mr. X. That is how it stays unless he instructs me to the contrary. I strongly suggest you or your salespeople follow the same rule. You can never go wrong with formal usage. If Mr. X has another form of address, then use it—for example, Dr. X. I've learned that older owners whom you contract with seem always to stay "Mr." Contemporary or younger owners seem to get around to first names as the job moves along, and I feel this is so because of the close contact necessary during construction. Even when Mr. X introduces himself as Jack X, you should forget the Jack and call him Mr. X. A little courtesy and respect will go a long way at this stage.

Why You Want to Talk with Mr. X

You, the construction salesperson, are always carrying the ball. You look like a construction person when you walk in, shake hands, and introduce yourself to Mr. X. What's next? Do you stand back and wait for the owner to cover you with information? Well, you'd better not. You get right down to business and let Mr. X know that you understand he might be making plans to build a new facility. If he is, then show him you're interested in his business. That's why you took the time to come in to talk with him. At this point, if Mr. X is at all an aggressive operator himself, he'll appreciate someone seeking him out and asking for business. I know this from personal experience, when someone shows some initiative by calling on me and the roles are reversed—as often happens with suppliers and subcontractors wanting to work with me on negotiated work. I find myself leaning their way because they had the get-up-and-go to come to me and ask not for the business but for a chance to earn the business. So I see no reason why our Mr. X won't react to a construction salesperson in the same manner.

Get right down to work. Find out if he's thinking about building, and if he is, make it clear that you want to know how you can help. This is your main theme, helping Mr. X. Another way to say help is to show interest. You show interest by offering to provide prices, get information on anything from drainage to construction loans, and check out any problem acting as a major roadblock to the progress of the project. It's up to you to convey all this to Mr. X while asking questions to see if he is a lead, and, if so, whether he can be qualified into a good prospect.

At first, Mr. X will be on the listening end of the conversation, mainly because you know why you're there. Mr. X doesn't, and it's up to you to tell him. Bear in mind that you always do tell this within the low-profile, soft-sell guidelines, at the same time taking note of Mr. X's situation during your meeting. Is he a wholesaler sitting in his office or a retailer standing behind his counter?

Continuing to Sell Yourself

Beyond this point you will have qualified the lead; and if he's a good prospect, you press on. You don't just sell yourself and do nothing else in the process. You sell yourself after the initial meeting by the way you carry out all the many steps of working up a proposal, and also all that takes place up until the contract signing. After the contract comes the construction phase; here again you keep right on selling yourself by the way you handle all the many details and problems for the owner. The rules for accomplishing this goal are fairly simple:

1. You do what you say you're going to do.
2. You communicate with Mr. X so he knows everything that he should know.
3. You follow up.
4. You do everything you can to take the load off the prospect.

The construction salesperson will have to do all four things—not two or three of them, but all four. It will mean a lot of hard work. The salesperson who is not willing to follow these rules while working with prospects is going to be surprised at the jobs he loses. He might have a large number of leads and prospects at first but few contracts; and as contracts become fewer, so do the leads because business stimulates more business.

Let's discuss each rule. First, do what you say you're going to do. Much of what you'll be doing for Mr. X will be at your own suggestion. You'll ask if you can work up some prices, or if you can take a crack at a drainage problem; next you set out to get it done, and then report back to Mr. X. Don't leave the prospect with, "I'll see what I can do and get back to you when I can." Tell him instead, "I have a couple of ideas to check out, and I'll be back in touch in a week. Is next Friday convenient for you?" Then the following Friday see him. If you find you can make it sooner, call first and arrange a meeting.

Sometimes it's impossible to get all the work done in order to call back when you're scheduled to do so, which leads us to the second rule: Communicate with Mr. X. When you can't keep an appointment, call and tell Mr. X. While doing so, don't just say you cannot make it. Tell him why: somebody was supposed to supply you with some information and didn't do so, and you're still waiting. Or maybe something has come up that has nothing to do with the job but is more important—then I recommend you come up with some reason that has messed up your schedule. Don't tell him that something more important has come up! You want him to think he's the most important person with whom you are working.

Communication will play a vital role during the construction phase. Keeping the owner completely up-to-date lets him look good to his friends and business colleagues. When they ask why his metal building that was due to be delivered last week has been delayed, Mr. X can come back with the full story, and then go on to tell his friends he knew about the problem three weeks ago. Keeping Mr. X informed is one of your best ways of selling yourself. Keeping the prospect up-to-date allows you to accomplish another necessary function—keeping your face in front of Mr. X. It's important for the prospect to see a lot of you. It shows him you are very much interested in his business. Communicating is one of the best reasons for letting Mr. X see you.

When stopping by to report some detail to the prospect during the selling process, it's not necessary to call every time. It's perfectly okay not to phone if you're only going to be there a very short time. If Mr. X keeps you for some reason, that's fine. But if he doesn't, get in and out in a hurry. The reason why you don't want to use the phone until you're on good terms with Mr. X is that he may ask you to tell him the information while he's on the phone, thereby defeating your purpose, which is to be seen.

The third rule is to follow up. This rule overlaps somewhat with the second rule. Nevertheless it's important by itself. Don't assume anything; check back and find out. When you assign a duty to Mr. X (remember trying to establish interest), don't wait to hear from him; check back with him. Often something will fall through a crack because each party is waiting for the other to make a move. The construction salesperson has to stay in charge, so check back and follow up. This behavior also shows the prospect interest on your part. The fact that the prospect doesn't call you when he has the information doesn't necessarily mean he is not interested, so don't wait around for Mr. X to call.

During the construction phase you may have to follow up behind someone who is doing work for the owner. Say the owner has some type of special equipment that goes in the slab but is not in your contract. Somebody has to take charge, and it had better be you. That's the only way to look out for the owner's interest and yours.

The fourth rule is that you do everything you can to take the load off the prospect. The main reason for this is to keep you in control. Except for the questions used to establish the degree of interest of the prospect, you take care of answering the questions. For instance, a parking problem is noticed; right off the bat you tell Mr. X you'll check it out with city hall and get back to him. Remember to name a time for getting the details to Mr. X. When all the little nitty-gritty problems start popping up, jump right in and tell Mr. X you'll take care of this and that while he minds the store. By now you should be able to see what this does for Mr. X besides selling yourself. You're making Mr. X dependent on you. That's important!

These four rules can be boiled down to one word: *dependability*. The prospect has learned he can count on you, and hopefully you've dispelled a cloud I think all contractors carry with them. Take a poll on any street, asking, "What's your opinion of contractors?" I'm willing to bet 90% of the answers will be, "They never do what they say they're going to do." Granted we are maligned unjustly most of the time. The public doesn't understand the hardships of contractors. Weather alone can make you into a lying bandit. The owner will think your excuse is baloney when you have not hit a lick on his job and the ground is still deep enough in mud to lose a small pickup, driver included, in the wink of an eye. He thinks you're doing something else more profitable instead of working on his facility.

At the same time, owners have been pushed around, and they never forget it. Anyway, you have to prove to Mr. X you aren't like the rest of the bunch—you're dependable; he can count on you. And once you sell him on this, *do not* drop the ball and let it be your fault.

Returning Phone Calls

Another thing you can do is return phone calls. It's amazing how many businesspeople operate under the premise that if the call is important, so-and-so will call back. To me it's plain common courtesy to return a phone call. A week seldom passes that I don't recall what a prospect told me several years ago when I returned his call: "It's a pleasure doing business with a man who returns his phone calls." I might add that a 14,000-square-foot warehouse came from that particular prospect.

Now the construction salesperson has to do more than return the call, be it from a lead, a prospect, or an owner during construction. The salesperson has to make contact and communicate. My point is that you don't just return the call and let it drop if you miss Mr. X. You keep trying until you talk with your party. The caller will be pleased you got back to him; it also shows him you're interested in his business. All of this helps sell you to Mr. X, which is the name of the game.

Selling yourself to the prospect successfully requires all of these:

- Good first impressions
- firm handshake
- Addressing Mr. X with respect
- Quickly explaining why you want to talk to Mr. X
- Doing what you say you're going to do
- Communicating
- Following up
- Doing all you can to help take the load off the prospect and make him dependent on you

By doing your best to adhere to these rules, you'll do a good job on selling yourself to the prospect. It is hard to determine how much the prospect actually buys of you; probably the only true yardstick is the contract.

A word about trying to sell and impress with social engagements. I don't do it, period. I feel that it's not necessary. I am a businessperson doing business with another businessperson. I leave it at that, and I'm positive the prospect or future owner appreciates having our dealings kept on a business level.

I'm not talking about having Mr. X join you for lunch so the two of you can get away from his phone. Entertaining may be an important part of selling in your part of the country; and if you feel it helps you sell a job, then by all means, go ahead.

SELLING YOUR COMPANY

You've now sold Mr. X on yourself; the next step is to sell him on your company. In other words, you have to prove to the prospect that you can do what you say you can do. All Mr. X has from you is a lot of words, and your words have not cost him a cent so far. But can you perform? The best way to answer this question is to show the prospect past jobs and tell him some of the details as well as for whom you have worked. There are several areas to be covered.

Pictures

We've all heard that a picture is worth a thousand words. In construction selling, it's true. Have with you at all times top-quality pictures that will show the prospect the caliber of work your company does. Snapshots will do, but larger pictures are more impressive.

Snapshots can be enlarged with no trouble, but there is usually a limit beyond which the pictures start losing their clarity. The handiest size for showing details is 8″ × 10″, kept in an attractive folder. I suggest that for really good pictures you contract with a professional photographer. I've used professional pictures and they are well worth the expense. The pro knows how to play light and shadow to highlight a particular detail or catch a building at the right time of day. The more impressive the pictures of your work are, the better job of selling you can do. It takes good tools to do a good job.

The secret to using pictures is not to overpower the prospect with them. After he's looked at four or five, it gets to be old-hat to him—in short, boring. Four to five should give a decent cross section of your work, which is all that is needed.

Besides their use in selling your company, pictures are great to move the job along, to get Mr. X thinking about details of his new building. This use requires more than the small thin folder used for selling the company. You need to have pictures of all different types of buildings and should break them down into categories, such as shopping centers, auto dealerships, retail shops, office buildings, and storage warehouses.

These pictures don't have to be of your jobs. They can show anyone's, though you should ask permission of the owner before you take any pictures. The same is true for the jobs you have built. I don't know of anyone who has ever objected, and it's a thoughtful courtesy.

Trade magazines are an excellent source of good pictures; don't hesitate to cut them out and use them. These pictures aren't shown to the prospect unless you feel they can be helpful. For example, while looking at the company pictures, Mr. X singles out one saying that's what he's thinking about, but with larger plate glass and a brick front. You know you have something close in your picture file, so you get it. It doesn't have to be exactly what Mr. X has in mind, only close; but it really can give the prospect a lift to see a picture of what he wants. I sold a tire store by using a metal building manufacturer's sales brochure that had a picture of just what the owner wanted. When I showed it to him, he said that was exactly what he wanted his new branch to look like. I had a signed contract one week later for a 7,200-square-foot retail tire store.

Showing Jobs

Putting the prospect in your car and showing him actual buildings your firm has constructed is another way to sell your company's ability. Be prepared for the prospect to offer some resistance to this idea. I don't know why, but it's always a big deal for Mr. X to get away for a while, even to do something that seems important. Remember, if he does take you up on being shown some jobs, it's a very good indication of interest. When you actually get Mr. X in the car, he becomes a captive audience; don't make him sorry he's in the car with you so that he can hardly wait to get back to his store. Do a little talking about the buildings you are seeing, but don't overwhelm him with construction details. He couldn't care less! He will want to hear about how much a building costs, how so-and-so is doing since he moved in, or whether that type of side-wall treatment is expensive.

Remember to do some listening; don't do all the talking. And try to show Mr. X what he's interested in; don't show him schools when he's thinking about a new warehouse. Don't keep the prospect away from his business any longer than you said you would—be very careful about this. If the prospect would like to see something for the second time or a building that might be a little out of the way, then go. And don't hurry him back because you have something else to do. If I have a meeting scheduled that may go into overtime, I never commit to another appointment behind the meeting. It's a very bad selling technique to have to cut Mr. X short when he's interested in talking with you. If at all possible, plan your time so there will be no conflicts.

The situation with showing jobs is the same as with pictures; they can be used to sell the job as well as the company. In showing a job you'll want to take the prospect on a tour inside if possible. And right here the bad jobs come back to haunt you—it would be something if Mr. X wanted to see the interior of the building you did last year and the job had been unpleasant at the end! I'll never be able to say it often enough: the happy owner is the best salesperson you'll ever have.

Okay, so you have an owner who will let you come back, so you arrange a meeting to show his building. Always do this. Don't barge in; that's bad manners. Here is where the car phone can be put to good use. When you and the prospect show up, go straight to the owner and introduce the prospect. You may be surprised; the chances are the owner will take over and conduct the tour.

Since you are in the construction business, you know from firsthand experience that no job is problem-free. Even you and the owner now conducting the tour probably had a couple of problems. The funny thing is, he probably will not mention them. He wants to give the impression that he was smart enough not to have any problems. Besides, he's flattered to be showing off his building. You shouldn't let some little problem stop you from calling on an owner to show his building. At the same time, don't be dumb enough to ask if you know he'll refuse. Why ask for grief?

A good salesperson lets his product talk for itself. You do the same thing by showing some of your past work.

Dropping Names

Be a name dropper. Every place has its prominent citizens. They don't necessarily have to be known to the general public, say as members of city council. They can be individuals who are recognized as leaders in an area familiar to the prospect. Use their names if you have done work for them. It helps put you a step ahead of the competition when the prospect knows your firm is accepted by the movers and doers in your community.

It helps as well not to hit Mr. X over the head with a name. Slip it in gently, and then let the prospect pursue the subject. For example, don't say to the prospect, "I built Mr. So-and-So's new facility." That might be taken as boastful by Mr. X. Instead say, "I faced the same problem, Mr. X, just last year, and the situation was taken care of after we brought in the city engineer. It happened to be on the job we did for Mr. So-and-So." Now you've told the prospect your firm has done work for a certain person in hopes that it will impress him; and you've done it without being too pushy. At the same time, you're telling Mr. X you can solve problems.

Along with name dropping, be prepared to give out references. The convenient way to do this is to have reference material made up that becomes another selling tool. (See sample on page 108.) Here you list several jobs, and I would suggest no more than five or six. That's enough to get the point across without overwhelming the prospect. Along with the job names give the address, phone number, and the person's name to contact. This keeps things easy and simple for the prospect to use.

This reference material is designed to sell your company, so include your bank and a name to contact, then the insurance firm with name, and a list of suppliers with whom you work. This reference material allows the prospect to check out your firm and determine you are what you say you are.

There will be some work involved because everyone on this list should be contacted and approval given to use their name and company. Also, here is where you can stress a very important point that is a part of design-build construction: repeat work. Whenever possible, let your job list have repeat customers. On the sheet I'm using all five of the projects are second and even third jobs I have done for the same business. This tells the prospect that your company is well accepted in the business community.

Keep the reference material to one sheet of paper with the company letterhead.

I also recommend using the negative reference. This is not a reference who will bad-mouth you. *Negative reference* is a term I started using several years ago, the idea being a holdover from my earlier days on the road as a factory sales rep for a building materials manufacturer. The old hands told me when I started selling to always remember that a well-handled complaint will impress your dealer more than anything else.

I put this idea to work in selling construction. While I'm giving out the reference information, I can usually get a quick double take with, "And while you're checking, Mr. X, call Mr. So-and-So. We had the biggest headache on that job." The prospect will be a little surprised to have this type of reference offered. I go on to explain that just like him, my company has its share of problems. The point I want to make with Mr. X is that what really counts is how the problems are handled. I make this very clear and invite Mr. X to call the negative reference and check for himself. I've found that such an offer carries a lot of weight and truly impresses the prospect. And the negative reference has been checked by the prospect from time to time. I believe, though, the offer is proof enough to Mr. X that you intend to shoot straight with him. It sells him on you as well as your company.

Company Brochure

I've had occasion to use an attractive professionally done company brochure, and it can be an excellent sales tool. You're able to accomplish several things simultaneously with the company brochure. You can present the whole company image in one neat package. Pictures of past jobs along with captions giving the size and use of the buildings plus the owners' names allow you to show pictures, drop names, and provide a ready list of customer references at the same time. You have the added advantage of being able to leave the material with the prospect for him to study at his leisure. It also keeps your name in front of him much better than a calling card.

I've had experience with company brochures, and there's one thing not to do if the pamphlet is going to be used as a sales tool: Don't show pages of employees' pictures and lists of high-powered personnel with all their qualifications. Show buildings; that's what you're selling, not people.

I've seen a brochure full of people backfire. The prospect was scared off because he saw the company was obviously carrying a very heavy overhead load and just knew it was going to be reflected in the price he paid.

Company Letterhead

XYZ CONSTRUCTION COMPANY

Address Phone
Fax

REFERENCE DATA

Note year when company started

Having been in business some years looks good to the prospect. However, if it's a short time, say a year or two, I suggest this be omitted because the prospect may take note and want a firm with more time and background.

Projects

List four, five, or six (no more, that's all you need) projects with phone number and party to talk with. This provides the prospect with positive information that your firm can do exactly what you say.

Credit

Here I suggest you list three particular areas:

1. Bank, phone number, party to talk with
2. Insurance, phone number, party to talk with
3. Suppliers and subcontractors

List five firms with names and phone numbers. The credit references let the prospect make his own determination that your firm is a respected member of the business community.

Selling yourself and selling your company are really rolled up into one big selling process. It's very difficult to separate the two completely and concentrate on one at a time. Time is one problem. Not much time is required to sell your company; after Mr. X accepts the fact that your firm can indeed do the caliber of work he needs, then that's it for that subject. On the other hand, selling yourself is an ongoing process with no actual stopping point. You have to be able to jump back and forth in these two areas.

This is all fine, well, and good, you say. But how do I know when to stop this type of selling and get on with selling the job? The answer is: I don't know! Prospects are all different in some ways and alike in others. It is strictly a judgment call on the part of the construction salesperson. You might call it a feeling, a change of attitude by the prospect. The successful salesperson is constantly on the lookout for such changes, but don't expect them to hit you over the head. More than likely, it will be an innocent question. The best yardstick I can offer is interest on the prospect's part. Questions will be the indicator to watch for; and when you think the time is right, get to it!

SELLING THE JOB

The very first impression you want to give to the prospect is interest. You bear down on all the requirements and stress to Mr. X you need all the information possible if you are going to do a good job of helping him. *Help* is the key word. Don't start with, "Give me all the information you have." Instead start with, "In order to help get this project off the ground, Mr. X, I'll need some facts concerning first the site and then the building size." Don't throw the subject open for debate. You keep control. You ask the prospect particular questions, and you'll be pleasantly surprised how smoothly the interview goes. The prospect will be really starting to think about his new building, and the excitement will be there.

While you're working on the requirements, please remember you will not learn a thing by talking. I mention this for a special reason: As you gain more experience you'll come to know firsthand, if you don't know already, that there are certain places during the selling process where long periods of silence develop—for example, when the prospect is thinking about whether he wants an 80-foot-wide building or a 100-foot one. Your questions will make him think about the building, so *let* him think. Many salespeople believe that every time a period of silence occurs they must immediately start to chatter. Fight the urge. It's hard, I know, but better to learn now than the way I did—the prospect asked me to be quiet so he could think. It was a little embarrassing; however, I still sold the job.

A heavy silence will also develop when Mr. X reads your proposal. Accept it and sit there and think about the Super

Bowl, or where you're going over the weekend, or anything. Think; don't talk; let the prospect read. After the reading will come the questions, and then you can talk.

It's important not to tell the prospect what he should have. Just record his requirements and ask questions that add to the requirement knowledge. Keep in mind that you're trying to find out if the prospect has a good idea of just what it is he wants, and at the same time work up enough data to put together a budget price. Your recommendations can come later. This method works very well with the prospect who is starting from square one.

Let's take another tact for the prospect who had his plans and was stopped because of the prices that came in when the job was put out for bid. You've got a ready-made starting point, the plans. Again ask questions; display your interest and intent to help. Using the plans, proceed to point out areas where savings can possibly be recognized. Money is the attention getter, so sell savings and places where the cost can be reduced. Hopefully by now you've found out what the bids were and what the prospect has established for a budget. If not, then try to find out. One way is to come right out and ask. After all, the worst that can happen is that Mr. X will refuse to tell you.

You will not be refused very many times if you preface your question with the positive statement: "Mr. X, I know my firm can work up some redesign ideas to cut costs. It will be most helpful to know the budget we have to shoot for. Early on we should be able to have a good indication of how the reductions are running, and it'll save me from spinning my wheels if I know just what your budget is. I'm sure you don't want me wasting my time on a budget that will be hard to reach." You've taken a very businesslike approach, and for the most part Mr. X will respect you for it and give you the numbers you need. You may not actually need the figure to do your work, but this is a sure-fire method to obtain the numbers. The fact is, you'll probably really use the budget as a yardstick.

Think and Discuss Change

I've used this method many times to start a job moving along seriously. Don't be afraid to suggest a radical change to Mr. X if it will help accomplish a certain goal. An example may make what I'm saying clearer. Let's say our prospect is a car dealer, and he wants the showroom and offices in one building and his service area in another, completely separate structure. You work your budgets, and the price is too high. At this point you tell the prospect the price can be reduced if he puts everything under one roof. This is a radical change from the original requirements, but you are showing Mr. X how to reduce costs. You're doing your job, and you're helping. He may say no; and if he does, the prospect will have to revise his budget. Another change could be relocating the building on the site or making changes in the site itself.

This tactic is used with the prospect after you've worked up a price, all the requirements have been worked out, and the price is too high. You use the radical change right off the bat with the rejected-bid prospect; it is especially effective here. In this particular case the preliminary work has been done for you.

When negotiating with a prospect who has plans, always take the plans with you and as many copies as you can obtain. Sure, you're thinking, how else will you be able to work up the needed information? Think about this one, though. What you have no one else can look at! Right? So take those plans out of circulation, and keep them out as long as you can. Usually the prospect will keep a personal set, and there's nothing you can do about it. He probably won't let them get away, so the competition still has nothing.

You don't have to be a first-class construction superintendent to be able to suggest radical changes. All you're trying to accomplish at the outset is to keep the project moving. Don't even try to quote prices. Explain to the prospect that you'll have to work on the details, and you'll need the plans for a short period. Later the real construction people can tell the salesperson if indeed there is a savings. However, if the salesperson does have expertise in this area, then great. It will make him more comfortable with this segment of the selling process, hence making him much more effective.

It is imperative that the construction salesperson know something about construction in general. The very first rule for a salesperson is to know his product. Remember that the radical change is an attention getter; once in a while the prospect will buy your suggestion, but more often he will not. But you have now put yourself right smack in the middle of the project. You're now part of the picture, and that's all you wanted. Any other pluses are a bonus.

Many times a radical-change suggestion will not be necessary. Small changes may do the job just as well—substituting one type of side-wall material for a cheaper one, for instance. Then there are the changes you recommend that will not necessarily save any money but will give the prospect a better facility for his particular use, such as a one-way mirror in his office so Mr. X can observe his sales area while working at his desk, and yet remain private. Or maybe you could work out a better traffic flow for trucks around his warehouse. All such changes sell the prospect on your interest in his project.

There are several suggestions that I've more or less standardized and fall back on when nothing else is obvious: (1) *skylights in storage areas;* (2) *windows in offices to allow Mr. X visual control* (this is a good term to become familiar with; it's one that retail and wholesale prospects will want to talk about, and by your doing so, you show the prospect you know your business); (3) *plans for future expansion;* (4) the possibility that the prospect may have room for *building some space to lease out* (which I use to a lesser degree).

Somewhere in these four topics I can usually find a subject for discussion, and I've never had a prospect who didn't seem to appreciate the suggestion. With a little experience you'll be able to look at a set of plans or take the owner's requirements and see immediately how to come up with some good suggestions.

There is one special item that I always try to sell the prospect on considering, where I personally feel I would not be doing my job of looking out for the prospect's interest if I did not. The main theme is using the building as an insurance policy. I tell the prospect the advantages of designing the building to have multiple uses. For example, Mr. X wants a 60′ × 100′ building to operate a retail business; a 10-foot side-wall is more than high enough for his use. I then take the role of insurance agent and ask him what happens if something should happen to him. At first Mr. X may seem puzzled by the question, but almost always the answer is that the building goes to his wife. Then I ask if Mrs. X will continue to run the business. Most of the time the answer is no; the chances are Mrs. X will sell or perhaps lease the building. Then I ask him the final question: Why not leave Mrs. X with something that's really worth something in the marketplace? Believe me, you'll get a quick response, and he'll ask exactly what you're talking about.

At this point, I explain to Mr. X that a 10-foot-high building, while working very well for his business, still limits the number of businesses that could use the structure. It would pay him to consider a 14-foot-high building, which would be more versatile in use, hence more marketable. I point out that he could look at the added cost as a one-payment insurance policy. Every time I bring this up, the prospect considers the suggestion, and I would say that in many cases the owner does make the changes.

I don't like to leave the prospect thinking about not being around, so I tell him it's also a good retirement plan for him. When he does decide to retire and maybe sell or lease the building, it will be more marketable and a good source of income. This statement is always received with a smile. At the same time, I increase the contract, which makes me smile.

Often the prospect will plan to build a single-use building, and the construction salesperson should stand ready to point out the pitfalls, from both a use and a financial standpoint. Bankers are a gloomy lot about lending money, and they always take the attitude that Mr. X will go under two weeks after he moves in, leaving them stuck with the building. They frown on a single-purpose building. There'll be times when a single-purpose building is what is called for, and nothing can be done about it (for example, a motor freight terminal).

Budget Price

You've been hearing about the budget price all through this book. We came across the budget price in Chapter 3, and you probably remember that it was something that had to be given out with no guarantees of getting something else in return. The budget price is just what it's called—a budget, nothing more. It's a handle for the prospect to grasp. It's necessary for the construction salesperson to understand exactly what a budget is, and, even more important, to know how to use it.

There's no way to convince the prospect to move beyond a certain point until he gets a price. Everything he does is aimed in this direction. He wants to know how much the new facility is going to cost. This is even more true with a rejected-bid situation, only the main thrust is how much under the low bid can you get.

The prospect gives out his requirements only to allow you to work up the price. You're not going to get down to the nitty-gritty of selling Mr. X until you present him with a price. Then and only then will he press on—or maybe fall over in a dead faint and forget the whole thing.

At times the budget price and the contract price are one and the same. This usually occurs with a simple uncomplicated job, where from the basic requirements you're able to quote a firm contract price. This price functions the same as a budget; it gives the prospect what he has to know, and in this case he's better off because he has an exact price. In truth it doesn't eliminate the purpose of the budget, only having to work up a contract price later. As I said, this applies mainly to the simple jobs; for the more complex ones the contract price can come only after plans and specifications are prepared for pricing.

In working up the budget price, there are many ways to save time and money. Based on your own experience with past jobs and the experience of your key subs, you or your estimator, if you have one, will be able to whip up a budget with no trouble.

The key to a meaningful budget is a decent preliminary drawing outlining all the requirements. We went over preliminary drawings briefly in Chapter 3. A simple line sketch is all that is needed when you are talking with a prospect. As you question him and take down his requirements on your sketch, be careful to note all the details possible. This sketch need not be a super-professional drafting job. It's easy enough for the shakiest hand to make a passable drawing when grid paper is used. The grid paper (1/8 inch to the square is the best size, I think) looks good and allows you to work to scale. Often problems will pop up when Mr. X's ideas are put down to scale, giving the salesperson a perfect opening to make a few problem-solving suggestions.

I can assure you that in making suggestions you do not have to be an architect to know what you're doing. It's really very easy; and after you've worked with several prospects, it will become even easier. With experience you should become good at it; in this area practice makes perfect.

I do recommend sitting down with subs in the plumbing, heating, air conditioning, electrical, and site work areas for a short crash course in their particular trades. It's important to know how duct work is run to service a showroom, or how the overhead lights will have to be arranged, or the most economical way to handle three bathrooms, and when you can substitute gravel for blacktop. I've never known a sub who wasn't delighted to tell someone about his trade. Also, when you have a special problem in a trade that takes some expertise, give it to a sub. Most of them don't get a chance to work on something really interesting very often, and they'll jump right in.

When I was building the three skating rinks for one owner, we had a real problem with the electrical system. The owner wanted all kinds of lights over the skating surface, which could be manipulated in dozens of ways to obtain certain effects while the people were skating. In addition, everything in the entire building had to be controlled from a control center so that the manager could sit behind his console just like a disc jockey and have control over the rink. And the electrician had to coordinate his work with the people putting in the audio system because everything came back to the same central console.

I arranged a meeting with the electrical contractor and the owner, and the three of us sat down, told the electrician what we wanted to accomplish, and turned the project over to him. The man did a super job and really enjoyed putting the entire system together. Use your subs because they can provide backup for your budget. I've already stated my feelings about shopping a sub after he has helped you nail a job down, but it's worth repeating: *Please don't do it!*

After I have worked up the budget price, I add between 5% and 7% to the total amount to make sure I can build what the prospect wants within the price that I state. Always explain to the prospect exactly what the budget price is. It's a not-to-exceed figure for him to use. I always stress the point that it would be very easy to "low-ball" a figure, get his name on the dotted line, and then start with the added costs. This statement accomplishes two things, I hope. It lets the prospect know I'm trying to do the best by him I can (selling myself), and it puts him on notice to watch out for this maneuver from the competition.

The way the budget price is presented is just as important as the number. All prospects have some figure in mind before you give them yours. You may never know what the figure is because it may be so low, as it is in most cases, that the prospect is embarrassed to mention it. The point is, he will have a figure. It will be based on what a friend built the same kind of structure for last year, or an industry figure put out by some national headquarters, or what he built his last warehouse for 10 years earlier, with his idea of cost increase added in. Therefore, the prospect will have some idea almost every time, and chances are he'll have it broken down to cost per square foot.

You've all heard the old adage that "A little information can be dangerous." Well, it's never been more true than it is with the prospect with a square-foot price based on goodness knows what. The real hooker is that when he talks to other people, they won't always tell him the whole story—and this, I feel, is not done intentionally. The problem is that the owner confuses building cost and total construction cost. When asked what his building cost is, the owner will quote the building cost, but leave out the site work and all that goes along with it. The prospect then takes this number times his area, and presto, he has what his new facility is going to cost in total.

There is another source of semifalse information for the prospect, and that is the owner who feels he paid too much, and rather than take a chance on looking foolish in the business community, he will deliberately put out a lower figure. One way the embarrassed owner can answer the question and not look dumb, he feels, is to give the building cost only. In any case, the prospect still ends up with a dangerous number, which the construction salesperson will have to overcome.

When the salesperson presents his figure, it must be broken down into at least two segments: building cost and site cost. Never give the prospect a total turn-key price. I'm speaking from bitter experience. When I started selling construction, I worked with a turn-key price, and I was losing contracts at an alarming rate. It took me six months and the help of another salesperson experiencing the same problem before we were able to solve the mystery. Our competition was quoting building cost and site work separately. Many times the site cost was never priced, only the building cost. The building cost our competition was quoting most of the time was somewhere in the ballpark the prospect was expecting, and when he compared our price to that of the competition, we were promptly forgotten.

You may ask right here, what's the big deal? Tell the prospect what your price covers, and he'll be able to make a fair assessment. I wish it were that easy. Let me digress a moment and tell you a not-very-flattering fact about prospects and leads as well: they only hear and retain 10% of what you tell them!

When I follow up with someone after talking with him the month before and am told he has signed with a competitor, I don't mind it—that's business. But it really gets to me if he then says with a surprised voice, "I wish I'd known you did this kind of work." The fact is, I did tell him the month before; it just didn't register with the dummy. This to me is one of the most frustrating aspects of construction selling, one that sorely tests my objectivity. You have no way of knowing if you are really talking to the prospect or just speaking words that go in one ear and out the other. As you gain experience, you'll encounter this type of thing, and there's absolutely nothing you can do except grin and bear it.

Thus, you can't explain something as important as the budget price and expect the prospect to remember how to use it. At the same time, it's confusing to the prospect to try to study the situation when he has to use two systems, especially when he knows nothing at all about construction. You cannot teach him to use your system when he has one he already understands. You have to make your information fit his format, and not expect the prospect to learn to use something new to him.

After the other salesperson and I discovered what we were doing wrong, we changed our budget price structure, and the contract-signing rate started improving immediately. Lately I've found the building and site price have to be separate, for with all the new restrictions imposed by city hall, it's almost impossible to come close to a site price without an approved site plan from the city officials. I imagine you have found the same thing where you operate—more and more added to site work by city hall that really drives up the cost and in some cases puts the project out of sight for the owner.

When presenting the budget, push the building cost and stay away from the site details. Prospects have a way of thinking site work will cost what it will cost, and there's nothing to do but pay; so capitalize on this attitude and stress building cost and places to save. Nevertheless, chances are that the numbers you give the prospect will be more than what he has in mind. You'll know this from a long, drawn-out whistle or in some cases from a dead silence that lasts for what must seem like a year. He's trying to absorb what you've just told him, so keep quiet. He'll speak when he has his breath back.

After Mr. X has the price, you're going to have to start getting him used to the figures. A trick I use here is to put the ball back in the prospect's court, so to speak. I ask him what prices are doing in his field, and can he buy items today for the same price that he was paying five or even ten years ago? Naturally the answer will be no, and you tell Mr. X that you can't either. The two of you are in the same boat; you just sell different merchandise.

Usually at this point it's possible to obtain the prospect's budget. Simply say, "If the price seems high, Mr. X, tell me your budget and I'll see if it's possible to reach it and stay within your requirements. I'll have to make some changes, I'm sure, but there has to be a point we can reach to move along." Once you know what he has in mind, then you can go to work on any changes that may reduce the price.

Mr. X is now waiting for you to come back so he can see if he'll be able to build or not. He is becoming dependent on the salesperson. The salesperson must work at having this dependency increase by handling details and problems for the prospect every possible time. I'm not going to cover the numbers that go into a budget price. We are concerned with selling, not estimating. Besides, every locale is different, and only you know what's going on where you operate. My only recommendation is to put in a little slush for contingencies.

There is one thing to guard against when working with the budget price, and that is the prospect who wants you to break out sub prices. Watch out! He's shopping your prices. When this comes up, I tell the prospect I'll be glad to give him a price for the structure only, and he can contract with other trades directly. This does happen fairly frequently when you're selling metal buildings. The prospect will drop back to a structure price mostly to save money. There is one exception—when the prospect wants to handle a particular trade himself. Say his brother is a plumber, and he wants him to handle that area; then I will break out that price. That is the only time I will, though.

So beware of the prospect who wants a breakdown. Tell him what you will for an excuse, but don't give it to him. I handle this particular situation by telling the prospect that the prices are worked up on the past history of other jobs by the square-foot cost, it would take a great deal of time to work up sub prices, and it would not be good business to do so at this point in the negotiations. But here you want to tell the prospect nicely that you won't do it. After all, he is a prospect, and you don't want to burn any bridges.

Let's talk about an area where the construction salesperson does not feel very comfortable: the use of drawings with the prospect. Here's where the gray area becomes scary. The salesperson loses some of his control. You draw a sketch recording Mr. X's requirements, work up the price, and present it to Mr. X. At the same time it's difficult to explain exactly what the price covers without using the original sketch or maybe a more detailed one you used to take sub prices. Some design-build contractors offer drawings with their proposal price. I think this is a little risky, though. In any case, the prospect will want to see what the money buys, so a drawing is always part of the budget, if for no other reason than to help explain the details. It's supposed to be a simple sales aid, nothing more.

The problem arises when the prospect asks for a copy of the drawing, for any number of legitimate reasons—to show his banker or lawyer, or maybe to talk over with his spouse. Anyway, he wants a copy of the drawing.

I know you can see the trouble that might arise if you do

as asked. With the drawing Mr. X has all he needs to start shopping prices. The construction salesperson at this point is as vulnerable as he'll ever be. He has done a whale of a lot of work, and the prospect can take it and go on his way with it.

The only defense you have is not to leave a drawing if you can get away with it, and sometimes when I feel the prospect might be a shopper, I don't even put the budget price in a typed letter. It's jotted on a scratch pad, and that's what I hand to Mr. X. If he tries to shop it, the competitor doesn't know if it's a price from a contractor or something that was made up. A price on your letterhead is proof to your competition.

Don't work under the mistaken idea that the prospect won't show your prices and drawings to a competitor. He will! I'm speaking from experience. I've had some of the most beautifully prepared proposals handed to me, without asking. When you're talking about large sums of money and the prospect sees an opportunity to shop a few thousand off the price, he's going to do it. There's not much you can do about it except do such a good selling job on the prospect that he really feels more comfortable dealing with you and will let you have the last look. For this reason the salesperson never stops making the prospect dependent on him.

Back to the drawing. I try to combat this situation of having to give out the drawing by using an original one while presenting my budget proposal. It doesn't have to be the one you started with; it just shouldn't be a copy. While you're explaining your price, use the drawing to point out details and always have several questions about some particular detail. If there's a change, note it on the drawing. If no change, note that also. The main purpose is to mark up the drawing.

When you've completed your meeting and are gathering your papers, take the initiative with, "Mr. X, this is the only copy of the drawing, and we've made notes all over it, so I'm going to have it redone and brought up-to-date." I've never had a prospect argue with that approach. But nothing is completely sure, and every now and then a prospect will ask to make a copy for his records.

There are still times when, having left a copy of the drawing, I walk away with a slightly uneasy feeling in my stomach. It comes with the territory, so get used to it. Do everything possible not to leave a drawing because you're helping the prospect become a little less dependent on you.

I know of design-build contractors who work up a full-blown proposal with simple plans of the proposed project and give them to the prospect as a matter of course. I don't recommend doing this sort of thing at all. It has to run into money, and the prospect is getting way too much for nothing.

Salesperson's Exposure

This subject will overlap somewhat with the budget-price topic, so we'll jump back and forth at times. Think back to my statement that the prospect retains very little of what the salesperson tells him, as well as his impression of the salesman. The only way to combat this problem is exposure; in other words, Mr. X has to see a lot of you.

When you are working with a normal prospect who wants to build a normal building, say a $300,000 contract, and there are no exceptional problems, the time span from first call to budget price should be around two, maybe three weeks. I know this period will often vary, but I want to give you a feel for exposure because it's important. Why take so long to come back with the budget price, you ask? A few days should be enough. That's right; a few days are enough if you only want to drop a price on Mr. X and leave to wait for his phone call. But that's not what to do to sell construction successfully. The important thing here is to use the time interval to give you exposure to the prospect.

This period between the first meeting and presenting the budget price is what I call the work-up period. During this time you should make at least four contacts with Mr. X. Two or three should be in person and the others by phone.

It's important never to give out everything you know at the first meeting; always hold back something that you need to know. Even if you won't be able to put the price together without the data, don't be too concerned. Preparing the price should not be very time-consuming, so you'll have time to pursue exposure and put together the price.

At the same time there usually will be questions that should be put to Mr. X; but if there are none, then you'd better get busy and think of some to ask. You'll have to be the judge of what to see Mr. X about in person and what can be handled over the phone. My guideline is, how important a question is it? To me a question is important if you cannot work up the price without its answer. Everything else, of course, is less important.

Please don't bother the prospect with obvious little questions that make no difference at all during the work-up period.

They simply detract from the image the salesperson is trying to establish with the prospect. Very often during the work-up period you'll have to go back to the prospect for a command decision. For example, Mr. X is undecided whether to build a 50-foot-wide store or one 40 feet wide. You offer to help by telling him if there would be any savings, and, if so, how much. I'm not suggesting that you price two different buildings. You provide Mr. X with just enough to make a decision, nothing more. It's really not too much of a job to determine that there would be a savings in glass. Anyway, the salesperson will have to have a meeting to present the data so the prospect can make a decision or choice. Then after the building is determined, you can go on with working up the price, at the same time making sure you give yourself as much exposure as possible.

I've found that solving small problems that really don't affect the price is an excellent way to stay in contact with Mr. X during the work-up phase. For example, is the water and sewer hookup at the back or front of the site? Or, will the prospect be able to put in a 40-foot driveway entry without special permission from the city? All of these types of problems can usually be taken care of by phone, and then you have a very good reason to contact Mr. X with the information he's waiting for. All of this activity helps sell you to the prospect.

Remember, it's very important to have all the exposure possible. You want the prospect to begin thinking of you as *his* contractor. During this time the salesperson can establish some rapport with Mr. X, which will go a long way when you get to the prices. It should make it a little more comfortable for all concerned.

Always have a reason to talk with the prospect. Never just stop by to pass the time of day. Mr. X doesn't have the time for this and neither should you. Have an obviously acceptable reason for calling on him and never say, "I was in the neighborhood and thought I'd stop by." No matter what the reason for calling on the prospect, you don't ever want him to feel he's being neglected in the slightest sense. You might actually be in the area, but never let Mr. X know this. You want him to think you made a special effort to talk with him about his project. Again you're selling yourself.

While it's most important for the construction salesperson to gain exposure, care must be taken not to overdo it. In other words, don't become a pest and start to worry the prospect. The salesperson must exercise mature judgment. I'm at a loss to pass on any meaningful guidelines. There are so many little things that make up the prospect's personality, and the salesperson is the only person who knows how to handle our Mr. X. By now you should have some idea of what you can and cannot do, so it's up to you to determine the best approach and be able to keep selling yourself without pushing his *off* button.

The salesperson must learn to be attuned to the little things that will tell him to back off a little. Maybe the prospect is a little short with you when you contact him, or it could be a put-off on a meeting. He may suddenly be hard to reach by phone. The salesperson has to be able to read the signs.

A salesperson friend years ago told me a short story about reading signs. He said a salesperson is exactly like the Indian scout of old. Both operate mostly alone surrounded by the unfriendlies, and the only way to survive is to be able to read the danger signs left along the trail. You know, the analogy is quite true; the successful salesperson is one who can note the signs, both good and bad, and can conduct himself accordingly. There's no book you can learn this art from; only studying human nature and experience will sharpen these abilities.

Timing

Timing in any sales endeavor is critical to success. It is doubly so in construction sales. Think back to my example about the toy man who decided not to build because he was having a bad day. I'll always feel I dropped the ball with that prospect. I waited all day to take the contract to him when I should have been the first person he saw that morning after he opened his doors.

The first rule of timing is never to make a price presentation if the prospect is in a hurry. You'd better recognize the problem and tell the prospect you'll come back when he's not so rushed. Mr. X will only hear a portion of what you have to say under the best of conditions; if his attention is elsewhere, he won't hear a thing.

Realize that when you call on the prospect he may be under pressure. Pay attention when you enter his place of business. If the pressure is obvious, then get his attention and tell him you'll come back or call later. Remember you will always take second place to his business, so don't get bent out of shape when the meeting you've rearranged your schedule for is called off. Once I tried for four weeks to pick up a prospect and show him a job that he wanted to see. Every time I stopped by his place it was bedlam. I would wave and tell him I would catch him later.

After a couple of times I knew the prospect was becoming embarrassed about the situation, so I went out of my way to assure him that the canceled meetings were causing me no inconvenience. He seemed to appreciate my being so understanding about his problems. And I did sell him a 4,000-square-foot building.

The second rule of timing is to try not to leave a decision to be made over the weekend. For example, when you have a price to present, and you need to get Mr. X's name on the dotted line, don't give it to him on Friday. Invariably the prospect will take it home over the weekend, and that's when the trouble starts. He has time to talk to his brother-in-law, who always knows of another contractor who he's sure can better the price. Think back to the example of the car dealer who told me on a Wednesday to put the paperwork together, after which, for reasons beyond my control, the sign-up meeting couldn't be held until the following Monday. This project fell apart over the weekend. I even hesitate to give the prospect a decision on Thursday; it has a way of ending up on Friday, then Monday.

A few years back a stockbroker friend told me about some customers who played the market daily and always sold before the weekend; their theory was that the whole world can collapse on Saturday and Sunday and affect their investments—just a point of interest to show that construction salespeople are not the only people who recognize the negative results the weekend can play on a business deal.

Momentum is as important in selling a building as it is in a football game. When the players have momentum built up, it shows in the way they march downfield rolling over the opposition. But let them stumble one time, experience a quarterback sack or an offensive penalty, and you can almost see the momentum walk off the field. The same phenomenon applies to construction selling. There is a certain driving force that the prospect gets caught up in, which makes him want to see his new facility a reality. This momentum will carry over after the contract is signed. Seldom does the prospect wonder whether he has done the right thing and try to get out of the contract. He may have passing thoughts that he keeps to himself, but once he's signed, he's committed; and he wants to keep the ball rolling.

But let the prospect get sacked by a blitzing linebacker and he'll lose his momentum. And that's what the weekend does; it allows the prospect to sit down and think about the money, the responsibilities, the problems, and, much worse, have them pointed out by someone else. The net result is a loss of momentum. This doesn't necessarily mean the prospect will not sign the contract. Once in a while the weekend reflection period makes the conviction to do the job that much stronger, but that's the exception. Usually the salesperson will have to start all over again selling the prospect—maybe by assuring him the problems really can be solved or perhaps in the form of a change. The prospect might decide to start out with a smaller building than originally planned. In any case, the construction salesperson should be prepared to "hold the prospect's hand" for a few days and work like the devil to get the momentum going again.

What happens when it's not possible to keep the prospect from having the proposal over the weekend, you ask? Nothing, is the answer. It's a dilemma that comes with the territory. The salesperson's only defense is to time it so the prospect will get the proposal on a Monday or Tuesday. I use a tactic that will be helpful to you. I leave the final day I'll give the budget price to Mr. X tentative, which gives me a good reason to make that last exposure call. If you're thinking that's not much of a reason to call on Mr. X, just to tell him you'll see him on a later date, you're right. So I give the reason for the call substance: I tell the prospect one of my sub prices came back too high, and I'm in the process of checking it out and will be back to him next Monday. That's what I call a three-in-one shot. I timed the proposal presentation for a Monday, gave myself exposure, and let Mr. X know I'm very much interested in this project by telling him I'm doing my homework with the cost.

Even when you manage to present your proposal to the prospect at the beginning of the week, there's no guarantee you can have it returned to you before Friday. This is a difficult thing to control. Nevertheless, try to control it as much as possible, and then follow up to get the answer before the end of the week.

Watch your timing, and try to catch the prospect when he's in a receptive mood. When the prospect tells you he's ready to sign, drop everything and strike while the iron is hot. Don't delay any longer than is absolutely necessary.

Often you'll have to prepare a letter or even a contract for the prospect to sign. This happens when changes have been made to the budget price proposal and it can't be used; or it may be that the prospect never had a formal letter from you. In any case, you need something for him to sign. So you have it typed up at once. Pay overtime or anything else you have to, but do it. Then with papers in hand, go straight to Mr. X.

You don't want to give the prospect the impression you're falling all over yourself to get his signature. It simply doesn't look good for the professional, so you cover the haste up with looking out for Mr. X's interest. Usually negotiations are somewhat drawn out, and prices from subs and suppliers can become a little shaky later on. At some time during the negotiation you should have mentioned the importance of nailing down prices. That lays the groundwork for you to hustle and get the contract signed. You're moving fast to secure firm prices as quickly as possible, and that's why you rushed right over with the contract. If a preengineered metal building is involved, you have a perfect reason—to get the building in the production schedule so that the waiting period is lessened.

Proper timing is necessary for a smooth job after the contract is signed also. When you need a decision, try to present the situation to Mr. X when the prospect is on the upswing. For instance, you would like to sell the owner on a change order that will benefit him but also put a little more money in your pocket—let's say, change the windows to insulated glass, or upgrade the heating plant to take care of future expansion. Approach the owner on the upswing, when positive events are taking place, such as the slab being poured. My favorite time is when steel is going up and the building is taking shape. I think you could sell the owner almost anything when he's looking at the roofline against the sky.

Of course, you have to take into account when the work has to be accomplished and keep any changes in their proper order. But whenever possible, sell on the upswing. This also goes for bad news, which in the construction business always seems to center around time schedules.

You must move to get the prospect's name on the line, and cover your haste by telling him you're saving money and/or time. Timing is extremely critical, so learn how to make it work for you.

Helping with the Financing

As I said earlier, you'll be astounded at how little the prospect will know about money. Oh, sure, he knows everything there is about his particular field, which mainly boils down to buying and selling. The monetary mechanics of this process are not at all complicated when compared to financing a new facility.

The first problem facing the prospect is that bankers and mortgage people have their own jargon. So do people in every other field. People in their own field expect the outsider to understand what's being said to him. The problem is it's often gibberish to the prospect, and he's somewhat embarrassed to say he doesn't quite understand all there is to know about take-out commitments, placement fees, or exactly what a construction loan is and how it differs from the permanent mortgage; or in some cases, if the prospect is going to be leasing space, there may be a rent-up platform to meet. Then you can get into interest rates, time frames, and tail-end balloon payments. And there's much more to utterly confuse the prospect, who only wants to build a new building so he can sell more paint, tires, or whatever.

Add to this all the paperwork required, and you think you're dealing with a government bureaucracy. Let me repeat what I said earlier: If you're not comfortable with the financing aspect of selling construction, then I suggest you take the time and effort to become knowledgeable in this area. I'm sure you know a bank or mortgage company official; ask him for some help, and he'll probably flood you with reading material. It's a very good idea to stay up-to-date in this field. Learn what's new and how you can use it. Keep up with interest rates and who's lending money at any given time.

Being able to talk over these matters will be invaluable in making yourself useful to the prospect, and that's the name of the game. You must be careful, though, not to talk down to Mr. X if you see an opportunity to help with the financing. You should be as comfortable with the jargon as any banker, but don't come on like one. The idea the construction salesperson wants to impart is that it's he and the prospect against the other guys.

I've learned that the best way to approach the subject with the prospect is to offer to provide him with information. It's important to give the money lenders a professional package with all the necessary data about the project at their fingertips. I touched earlier on what I call my "bank package," and now it's time to look at it in detail.

When he goes after money, the prospect is in the selling game as much as you are. He has to sell the lenders and prove at the same time he's good investment potential. One of the most effective ways to do this is for the prospect to show the loan people he's dealing with a first-rate construction firm with a track record. Money lenders are cold, hard realists; the uppermost question in their minds is, how do we pay off the loan if something should happen to Mr. X or his business? This, of course, is only one of many questions that must be satisfied, but it nevertheless is a key concern. The idea is that your bank package will show the lenders exactly what their money will be used for.

When I work up a bank package I include three parts: a cover letter explaining the requirements as given to me by Mr. X and stating the price, a description of the work to be done, and the drawings. I always point out that the price is a not-to-exceed budget, but nevertheless still a budget. Then I explain that a firm contract price can't be given until architectural drawings are prepared and priced, if that is the case. In some cases the job is uncomplicated, and I can give a contract price; then I state in the letter that we are dealing with a firm contract price. In all cases, architectural drawings will have to be prepared and put on file with the bank for reference during construction.

The work description includes a specifications list detailing what material is to be used where. This list doesn't have to be a long super-detailed set of specifications. I hold mine to around two pages, sometimes three. Remember, we're dealing with lending people, not construction people, so the list should be kept simple and easy to read. They will not be interested in the size of the re-bar and thickness of the concrete in the foundation, only that it will be designed by a licensed architect or engineer and constructed to the design.

For the drawings, one will be a site plan showing the building location—and I don't mean a site plan designed by an engineer that you can build from, but only the size and shape, and exactly how the building is situated. The second sheet will consist of a floor plan and one or maybe two elevations. The size of the drawings is important. I know it's easy to reduce them to 8″ × 11″ for convenience, but I feel it helps put your image across when you supply a larger-size blueprint, and I always use blueprints, not copies from the office machine. It's possible many times to have all the information put on one page, and that is convenient. But I'm after appearance as well with my bank package, and five or six pages plus two blueprints makes for a nice hefty booklet; it looks substantial. It's important not to furnish the blueprints separately but to fold and insert them in the folder cover. This keeps the package in one part, making it handier for office use. Along with my bank package goes my offer to accompany the prospect to the bank and be on call to explain any details that may crop up.

A bank package represents time and money, and you're reluctant to do all this without a contract. Well, so am I. The only time to become involved with helping the prospect secure his money is when you're protected. The way I handle this (here is where I'm getting ahead of myself) is with a contingency contract. I have a signed contract stating that if Mr. X can secure financing, I have the construction contract. I'm willing to gamble a little with the prospect, knowing that by doing all I can to help obtain the money I'm helping myself nail down a contract. I'll cover this again in Chapter 7 and go over another method to use for protection.

I do admit I don't go this far if the prospect is the least bit shaky. But by the time you get to this point in the negotiations, you should have a good readout on just what type of prospect you're dealing with.

Another problem to be aware of is that it takes time to obtain a loan. Money people never seem to be in a hurry, and you have to wait for them to act. This time lag causes a loss of momentum. It's to the construction salesperson's benefit to have one or two reasons to see the prospect during this period; it will help keep interest and momentum from lagging too badly. Finally, be prepared to lead the prospect step by step through the financial woods.

Build/Lease Options

From time to time you'll come across a prospect who has all the qualifications for a profitable job except that he doesn't want to tie himself up with mortgages and the spending of his capital, which goes along with financing a project. The new facility will mean more inventory, and more inventory means he needs more money.

Usually you'll find the prospect in a dilemma as to how to go about accomplishing his goal. His problem is that he doesn't just want a larger building; he wants one that he can work from efficiently, and in a location that's convenient to his trading area. There will probably be plenty of places to rent, but most of the time he will have to take something he doesn't want and lock himself in for years on the lease. The prospect can't see himself working and buying the building for the landlord.

The prospect ends up with a special set of requirements: having a building constructed for his particular needs on the site he wants, and paying rent for a few years with the opportunity to buy the building. Who can arrange all this for him, plus work out all the details that go with a commercial project? Our prospect has a hard order to fill. He knows general contractors will build, architects will draw plans, bankers will finance, and commercial real estate people will work out the lease details; but he's in the wholesale food business and wouldn't have the time to coordinate all these people to make the project fly even if he knew how.

Then in walks the construction salesperson, who explains to Mr. X that he'll take care of everything; all Mr. X has to do is sell food products. What do you think Mr. X's reaction to the salesperson will be? You're right; he'll welcome him with open arms.

If you're thinking this sounds too good to be true, you're dead right! The time involved is a killer. The salesperson is responsible for everything, and there won't be enough hours in the day, you'll think. The rewards will be there, though, believe me; I speak from experience: two 18,000-square-foot skating rinks, one 30,000-square-foot food warehouse, and one 20,000-square-foot moving and storage warehouse. In all of these cases the prospect wanted to keep his working capital intact and get into a new facility, and the lease design-build plan made it possible.

This type of prospect is located and sold exactly like any other prospect with one important exception—there will be a third party involved. What used to be a two-way, one-on-one situation now becomes a building triangle of tenant, owner, and general contractor.

As I said, you follow all the guidelines with the prospect. You still need the requirements, a budget price, and as many details worked out as possible, because lease design-build will necessitate the construction salesperson's selling the job twice—first to the prospect and then to the future owner. It's impossible to do so without doing your homework with the prospect. This provides the information needed to show the future owner how the numbers work out.

Along with the construction data, you have to arrange for the tenant and the landlord to meet and negotiate the lease terms. Don't assume you'll be present for this particular meeting; you may not be welcome. I always offer to be in the meeting, and several times I've been present; on the other hand, I have been politely told my presence was not needed.

There is one very important point that the construction salesperson must always remember: when it comes to the bottom line, you are working for the person who signs the check. This is sometimes difficult to do because you are in essence serving two masters. This arrangement at some time will lead to a disagreement; it may be just an insignificant trifle or a knock-down, drag-out. When you find yourself in this predicament, remember who you're working for. I'm bringing this up because I've learned that you seem to be on the prospect/tenant's side. I don't know why, except that you have built up a rapport with Mr. X, and maybe there is some empathy on your part.

After you sell the job or idea, you have to locate an owner. There are two main possibilities: first, you and/or your company can become the landlord; second, you can get one or more outsiders to be the owner.

Let's look at the first suggestion. If you and your company are in the market for some excellent real estate investments, then this type of enterprise can be very attractive. The property should appreciate with time, and hopefully there will be some cash flow as well as tax incentives. There is one place to be very careful—ask for and expect complete up-to-date financial information from the prospect/tenant. When the owner commits to the tenant, he'd better be sure the tenant can meet the monthly payments.

The second suggestion takes more of the construction salesperson's time, so you'd better be prepared. At this point "digging up" an owner is like "digging up" a lead. You have to know where to go and what to do, with one large exception, which is that you'll probably be dealing with some "serious players" in the investment circles in your business community. Locating these people so you may call on them will require a combination of many methods. Of course, the personal contact is one of the best ways. The way is up to you more or less, but one thing is a must—you have to go out and ask; they're not going to knock your door down.

A good starting point when looking for an owner is to contact people in the real estate business who handle real estate developments such as shopping centers, office buildings, and warehouses. If you bring them a tenant, they may be able to supply an owner in return. One word of caution—*security;* be sure you don't put your tenant on the street to be scooped up by someone else. Until you know exactly what's going on, play your cards close to the chest.

When you're dealing with a third-party owner, your situation is one-on-one, and you have a certain amount of control. It's also much easier to communicate and know where you stand. On the other hand, when you are dealing with a syndicate, it's no longer one-on-one, but one against a bunch. There will usually be one who brings up the question of construction cost and asks how they know the construction costs are competitive. I know what you're thinking: You take the deal to them, explain the numbers that have a profit for them, and someone asks about construction costs. You hand them a deal on a platter and someone is worried that you might be making some money out of it! Face it; there will be people like that. I've often wondered how such a person would react to my sending him a spec sheet for a physical exam or detailed legal work and asking for a quote. I'm mentioning this so you can cover yourself at the beginning.

As soon as I meet with the group, I tell them, in a nice way of course, that I'm there talking to them for the money I expect to make. I'm not putting in all the time and effort that a project of this caliber requires in order not to end up with the construction contract. Then I go on and explain the obvious: their main interest should be in their return. If this project clicks, and we all come to terms, I want it clearly understood that I build the building! This will take care of the price-conscious people, for they are listening to you for the same reason you are there—the money. I firmly believe the investors understand, appreciate, and respect this approach. The trick is not to handle yourself in a manner that pushes their *off* button. Remember, you're selling! And the sad fact may be you need them more than they need you. With the money they have, offers are coming to them all the time.

At this point the construction salesperson enters a gray area where he's exposed and vulnerable. So far everything with the investors has been verbal, and to provide them with the data necessary to form an opinion, you have to disclose all the information. You're hanging on a limb, and there's little you can do about it except push as hard as possible to put the project together. One advantage of working with a real estate broker who understands business ventures of this sort is that he should qualify the group before you have the meeting to lay everything out for them. Knowing they are interested ahead of your meeting will increase your chance of a sale. At least you won't be shouting from the rooftops for an owner.

There is another very good source of lease design-build ownership that should be checked out: firms that have money to invest and are looking for the chance to become the owner in a building triangle. I've personally worked with this type of firm, and their sole responsibility is to locate investment ventures and become the owner. It's a pleasure to work with them because they are knowledgeable about the field and understand the contractor's position, especially the fact that he intends to be the one to build. If you are not familiar with this kind of investing company, then check with people you know in the financial field—mortgage officials, bankers, and don't forget the stockbrokers. They should be able to help.

Let's assume the group of investors buys the package. You are now going to work with a group. They are busy in their own fields, and I've found that they don't want to be involved in the daily details. Sometimes one member will be the spokesperson. More often the real estate agent will act as the contact person with the contractor and the new tenant. I've found this to be a very good arrangement, mainly because you can usually get in touch with him with very little hassle.

It's important to show the outside owner or owners that you're just as much a professional as they are. Take the same pains putting together your presentation as you would a bank package. The bank package can be a good outline to follow, with the investment numbers added in; always have a copy for each party who will be present.

Your biggest problem will be time. You and the prospect are ready to get the show on the road, and the group isn't able to meet for two weeks. With one owner it's not too much of a problem, but with a syndicate it seems that one is always planning to be out of town when you want to hold the meeting. It will be a trying time, and all I can say is keep swinging. If time does start to drag out, push a little to get it moving; for the longer it takes, the more apt the prospect/tenant is to start having second thoughts, and loss of momentum can set it.

The build-lease package offers a very good source of profitable business for the construction salesperson if he makes the effort to learn about all the parts that constitute the building triangle. All it takes is some hard work and time. Ironically, selling the prospect is the easiest part of the triangle; locating and selling the owner or owners is the hardest part—and under no conditions should you ever forget that this owner is the party signing the checks.

Selling Parts of the Project

Whatever the reason, the construction salesperson sometimes will be faced with the price being too high. The prospect cannot or will not spend the amount that is required to give him what he wants. Operating under the principle that something is better than nothing, the construction salesperson has to be able to show Mr. X how to save money—and one of the best ways is to take the project apart, so to speak.

The first tactic is to suggest that the prospect contract out the site work himself. In your budget price this is separate, but the salesperson is planning to do the work. Explain to the prospect that with a site plan he can contract directly with the site contractor, and you'll be glad to supply some names. Often this offer is enough to eliminate the overrun. But then there will be the prospect who needs to have the price drastically reduced; when faced with this situation, I start selling the structure only. In the case of a preengineered metal building, it's a very simple matter to quote this part; and if you get the job, it's a fast off-and-on project. When you offer these options to the prospect, you must stay in control; the best way to do so is with the plans. You explain to Mr. X that along with your portion of the work, you will furnish all the necessary plans for him to contract directly with all the various subs. In addition, you'll be glad to recommend certain dependable firms for him to contact.

My experience has been that the prospect will use the people the salesperson recommends. This allows the salesperson to do a good turn for the sub, which will pay dividends down the road. Of course, if the subs you recommend were the ones you were planning to use, then so much the better. They get the work, and you have a rapport with the subs that makes for better job relations during construction.

Let's return to the budget price for the moment. We discussed the changes in price you had to supply for the prospect and why those changes are part of the picture. I'm bringing this subject up again to stress the point that many times these changes that necessitate new prices can and will be initiated by the salesperson. When you start dismantling a budget price or selling up, you're giving the prospect some new things to think about, and cost is very much a part of this package. So be prepared to do your homework and make the price changes, and also to offer new ideas along with the cost.

Overcoming Desire to Bid the Project

Early on when you start selling construction, you'll be faced with the prospect who is not sold on negotiating the job. He wants to put it out for bid so that he knows he's getting the low price. This objection is overcome by complete salesmanship.

The way I handle this problem is to attack rather than wait and end up on the defense. I tell the prospect at the outset all the advantages of negotiated design-build over the bid route. The key point is design-build. I push the fact that the prospect is going to get exactly what he asks for designed for his particular needs. This approach appeals to his ego to a certain extent and lets him feel he's really making things happen.

The first advantage is money. Some will say it is the only advantage, and maybe they are correct. I tell the prospect if he hires an architect and bids the plans, he will supposedly know the bottom price. The fallacy here is that he only knows the bottom line for that set of plans, and those plans may call for materials that are expensive and not necessary for a functional building.

I keep stressing that with the design-build concept the prospect will be able to know what his planned project will cost. If he goes out to bid, Mr. X will have to spend a large sum of money to have his plans prepared; then there is no assurance the price will be in the range he wants. At the same time, the prospect has very little control over what materials are used, and this has a direct bearing on the job cost. I ask such questions as: What do you do if the bid comes in too high and you feel it's impossible to build?

Then I go into my sales pitch. I stress control at all steps of the project. I'll supply the prospect with necessary information as to budget prices. Then when the architect is brought it, he'll be working for the contractor, not the prospect. This working arrangement allows a give-and-take atmosphere between contractor and architect that will enable the prospect to receive the best job possible for the money spent. The contractor and architect work out a set of guidelines. The architect then has a set of guidelines to follow when drawing up the project. The architect is paid to provide a technical expertise, but the design work has already been laid out by Mr. X. Both you and the architect are working hand-in-glove to please the prospect. The trick is to give him what he wants, and you as the general contractor can see that this happens if the architect works with you.

The second advantage I can offer is help. Numerous problems will arise, and I will be in a position to take the load off Mr. X. I talk about helping with the finances, supplying a bank package, and being present at loan meetings. I push past performance on similar jobs (that is, if possible), and how Mr. X can profit from this experience.

The third advantage is turning to my benefit the widely held belief that you can never get anything done when dealing with contractors. I hit hard that there is one overall person in charge and everything goes through him. I call this single source responsibility, and I keep driving home how it helps the prospect.

The last advantage I push is saving time. This should be important to prospects because they don't move until forced to, and then they are most likely operating with a deadline. I explain how time-consuming it is to have plans prepared and then put on the bid market. The prospect is always surprised to learn that an architect generally cannot turn around in a matter of days or even a couple of weeks. Then when I tell him it will be four to six or maybe eight weeks before he has the bid prices back in his hands under ideal conditions, he becomes concerned. The basic problem is the prospect has no yardstick to measure performance. He knows how it's done in his business and sees no reason why other people can't get results the way he can. Bear in mind how the retailer, the wholesaler, and to a certain extent the industrial businessperson operate. In a very broad sense, for these people to accomplish something, they simply take an item off a shelf and send it to the customer. For many, the time it takes to crank up a construction project is incredible.

On top of this time lag within the construction industry there's the frustrating tangle of red tape associated with having site plans and drawings approved in order to secure the building permits. I can only speak for my working area, but every year another regulation is added that makes getting the building permit more time-consuming. I'm sure you are finding the same situation where you work.

By now the prospect is concerned about whether he's going to end up in a time crunch. At this point I put all of these negatives to work for me. I tell the prospect the architect with whom I'm associated turns out plans in a fairly short time, usually within two weeks. There are two reasons for this, which I explain to the prospect—the details are all worked out. He takes the requirements and turns the information into plans. Second, I'm a steady repeat client for the architect, and thus get preferred service. It just won't do to tell Mr. X you're faster. Talk is cheap, so explain how you're faster. After all, the prospect is no dummy; don't treat him like one. If he ever thinks you are doing that, you can forget about the contract.

Next I stress how I handle the time problem at city hall.

Armed with the prospect's requirements and site information, I visit the various city departments concerned and determine exactly what they expect in the plans. This really is a time saver, since revising plans after they're turned down seems to take twice as long as it should. I especially do my homework on the site plan, for two reasons: time is the first; but the site cost is nothing but a grab in the air until you have some concrete facts to go on, and the city can provide them.

I'm selling myself to the prospect and at the same time showing him how I can get the job moving and keep it moving as well as supplying him with needed information. When I go to city hall, it's because I need help in some form. I simply ask for the help of the person with whom I'm meeting. Once in a while it's a sticky problem or interpreting the codes in the gray areas. I've never had a building official fail to try to be as helpful as possible when I've asked for help with information in solving a problem.

Back to our subject of selling the prospect on negotiating instead of bidding. After explaining how I can save time at city hall, I start to push saving actual construction time, mainly during the start-up period. When the job is bid, each part follows the one before it in a step-by-step process. Very little is done simultaneously; thus, a great way of saving time is not utilized.

The example I use is buying steel. Be it a metal pre-engineered building or structural steel, there is a start-up time period; so, I explain to Mr. X, while we are waiting for the permits to clear, I put all the steel on order. This allows the order to be in the production schedule, yet it can be withdrawn before a certain time if a problem arises and delays the project. In the case where a metal building is to be used, the building can be ordered before all the interior details are worked out, thus saving time. Impress on the prospect that you can save him time; and if he's under the gun, that will be most important to him.

Finally, I give the prospect my most important reason. I can save him money! Then I back up the statement. By knowing what construction materials cost almost on a daily basis, I'm in a position to recommend to my architect more economical ones. Also, my past experience on similar jobs will allow me to know what materials will give the best service for the money spent.

I know the prospect has some idea of what he wants to pay, so I push the point that working to his budget and saving time is more important than saving a little through bidding.

Overcoming the prospect's desire to bid requires the sum total of many suggestions, with the emphasis on saving money and time. Quite often you will be successful; then again, if you're not, you can always bid the job and add a redesign proposal that the prospect may find interesting.

COMBATING COMPETITION

So far you have been in the front-line trenches of construction selling. But how do you handle yourself when the enemy overruns your position and you end up in hand-to-hand combat?

You'll always have competition. Without competition we would all grow lazy. Competition ensures that this doesn't happen; it keeps us sharp and up to scratch. So instead of worrying about competition, welcome it—it's the force that makes the wheels turn faster, smoother, and for less.

When you are selling a prospect on not bidding the job, the situation in which he's taking proposals from more than one design-build contractor can help to satisfy his intention to check on the price and get a handle for the project from the marketplace. In short, he's sort of bidding the job. This situation goes a long way in convincing the prospect not to go out for a formal bid. It is the only case I know of where competition helps to make your job easier.

Even though other proposals give Mr. X a good idea of what the job should cost, he's still talking to your competitors. This is your next hurdle to overcome, and there's only one way: outsell your competition! You do this by outworking them, taking on and solving more problems, and outthinking them.

The first thing you don't do is bad-mouth your competition. A professional salesperson never does this. When you do, you are in effect telling the prospect he's dumb and stupid for dealing with such a no-good so-and-so Then guess what happens. Bang! goes his *off* button, and you might as well pack up and leave.

My standard reply when a competitor's name comes up is that it is a good firm and a good competitor. That's all. I'm not going to knock them, but I'm certainly not going to blow their horn either. The very first point I try to make with the prospect concerning the competition is to get him thinking about who's calling on him. For example, let's say I'm from a small- to medium-size firm, and my closest competition is a large outfit. Then I stress to Mr. X the good points about being small. I have less overhead; therefore, I am more competitive on price, with more personal supervision and interest in the project. I'm trying to have the prospect think how these positive qualities look when applied to the larger company. I want the prospect to start making comparisons between me and the others. Hopefully in this particular case he will realize the competition is indeed a large outfit, with all that overhead he will be helping to finance, and maybe he'll be concerned about not getting the individual attention he feels he should get.

This technique also can be used in reverse. If you're a large company, stress your good points, such as experienced workers, volume buying of materials, and better organization for faster construction. You, as the construction salesperson, will have to pitch your sales approach to bring out all the positive points. Then leave it up to Mr. X to draw his own conclusions. In my opinion, the size of your prospect's business has nothing to do with stressing small or large construction firms. The small businessperson may want a small contractor; then again he may like the idea of the largest outfit in town doing his work. There's really no way to tell.

Hard work will overcome the competition because they may not be willing to put as much into the selling as you do. If I had to pick an area where construction salespeople drop the ball, it would be *follow-up*. I have ended up with a number of jobs not because of anything great I did; I just kept going back and trying to make the prospect as dependent on me as possible. You might say that I simply outlasted the competition.

So many construction companies make one or maybe two passes at a prospect then leave him to call if he wants to do business. Think back to your being on the buying end; when everything is equal—money, service, and so on—how do you choose? Most likely the same way I do; you choose the person who shows he really wants the order. So do onto your prospects as you want done onto you. It's a sure way to beat the competition.

I've also found from experience that after the prospect receives his first prices and starts working with me, there'll be some changes that will usually add to the price, and these changes are seldom questioned; so I have a chance to improve my profit picture by a small margin.

Now let's look at what you do when you get into selling the prospect late. Your competition has already been working with Mr. X. The very first thing you set out to do after establishing the facts pertinent to the project is to muddy the water.

Your competition has a head start on selling and you need time to catch up, so you slow everything down by giving the prospect something else to think about. Let's say Mr. X tells you he plans to build a 100′ × 100′ warehouse; so you work up the price, and get back to him in a few days. During this time you use the regular ways I've gone over to gain exposure while the prospect is waiting for your price. You have probably stopped the job or at least slowed it down. Then when you do present your price to Mr. X, you throw in a hooker along with your presentation. In this case you tell Mr. X you'd like for him to give some thought to maybe changing the shape of the building from 100′ × 100′ to 50′ × 200′. You explain it will save some money, and the building will be easy to divide if he ever decides to lease part or all of it.

Chances are, the prospect will at least think about it. He may even call in the competition to make changes. And even if he decides against your suggestion, when the job gets back on track you're in the running. In a case where you're coming in late, pay attention when you're working on the requirements, with the idea of picking out some point that you can use to muddy the water.

There's no guarantee this tactic will work all the time. The prospect may dismiss your suggestion with a shake of his head and push on. The prospect should at this point know who you are. This at least gives you a fighting chance. Your best defense is to take the offense. Attack your competition. Do your best to impress the prospect with your special knowledge of construction. Show him a better way to get the job done. Don't be afraid to make suggestions. This is the second area where construction salespeople fall down on the job. They take the requirements as given and work from there, without a single thought on improving what Mr. X is starting with. Nothing else impresses a prospect like showing him a definite improvement, especially if it eliminates a future problem.

What you're doing here is *outthinking* your competition. A food distributor I built for is a perfect example. Mr. X wanted to build a new warehouse, and one of his requirements was to be able to back his 10 small delivery trucks into the building and leave them overnight. Early in the morning a crew would come in to load, or sometimes it was done at night. In any case, the trucks had to spend the night in the building. This also made the trucks secure from vandalism and thievery.

When Mr. X talked with me and my competition, he specified 10 overhead doors on the back of the building and explained how he intended to use them. Right off the bat I could see that 10 overhead doors were going to be 10 headaches, and expensive ones at that. Two weeks later I made my presentation and received no indication at all of what Mr. X was thinking. I suspected he had heard just about the same from all involved. He could have picked any one of us out of a hat; all he had to do was make up his mind. I didn't like the odds, so I spent one full day trying to come up with something different that would improve the job in some way and make me stand out. I kept going back to all those overhead doors. That was a potential problem area, but how could I capitalize on it?

That night at home it hit me: increase the building length by 15 feet and put one overhead door in each corner at the back of the building, but located in the side-wall instead of the end-wall as Mr. X had requested. This would allow the trucks to be driven into the building and backed up to their loading area. So now we had 2 overhead doors in place of the original 10, but at the same time we had more building. Cost now became the factor to consider. The next day I went right to work, and to my pleasant surprise and relief found that the price to Mr. X would not change. It was close to a wash-out because those 10 doors cost a lot of money.

I called Mr. X and asked for a meeting the following day, explaining that I wanted to show him a suggestion that would solve a future problem. I then spent the rest of that day working on a drawing to illustrate my idea. I even made several scale templates to represent delivery trucks. I must admit that I puzzled Mr. X when I called him that afternoon asking for the overall size of one of his trucks. He didn't know, but found out and called me back. Needless to say, I had his interest by now.

Right on time I walked into Mr. X's office, unrolled my drawing, and told him how my ideas would help him without a cost increase. To prove that the new layout would work, I used the templates to show there would be ample space to jockey the trucks about. After hearing me out, he led me out back, and we marked off the approximate distances of my proposed driving area. Mr. X then had a driver move a truck around the marked-off area. There was plenty of room, and I let out a silent sigh of relief. I had the job signed up before lunch.

There's no way I can give you a list of guidelines to follow that will enable you to outthink the competition. You'll just have to do it yourself; but if you make a point of trying, I'm sure you'll be pleased with the results.

I've often heard it said that timing is everything. I agree, it is, and very much so in construction selling. We touched on timing a couple of pages back. And I would like to bring up a special use of timing here. The case in question is that of the timing involved in trying to slow down a job so you can catch up with your competition. After meeting with the prospect, don't rush back the next day. Drag things out if you can possibly do so, but take care not to overdo it. The reasoning here is that a construction salesperson builds up momentum along with the prospect, and you want to insert a little slowdown into this momentum. If the prospect is serious, he will never really lose his, but it's amazing to me how many people selling construction grow cold on a prospect when there's a hitch. I think this goes back to my number one peeve, no follow-up. Timing is more important when muddying the water, but it must be handled with finesse!

A good way to use timing to combat competition is to be the last salesperson to make your presentation. If Mr. X is taking proposals from three design-build companies, and you are first or second, there is no way you can expect to get a decision from the prospect until the third proposal is in. In a small way the third salesperson controls the time when the decision will be made. The first two salespeople are being forced to conform to the third person's timetable. The contacts you maintain during the workup period can be most helpful with timing your proposal. It's easy during a conversation to drop some hints and find out when the competition is turning over their proposal. Ferret out such important information and use it to make sure you are last.

Keep in mind what we talked about earlier about presenting your proposal. Try to keep that time near the beginning of the week. So now you have two timing guidelines to follow: be last, and do it early in the week. When you're last, Mr. X can't discuss your price with the competition. It will be impossible to handle this situation as you would prefer every time. You can only try, but with a little experience under your belt, it does go smoother.

My last comment about the competition is that just about everybody is a nice guy like you trying to make a living. As in most business areas, everyone knows everybody else and has a fair idea of what's going on; thus, it makes sense to be on pleasant terms with your competition. This can be very useful to you when selling construction. The one kind of prospect that the construction salesperson has no defense against is the joker who is using you for a source of information and drawings if he can get them. This is the gray area where your backside is hanging in a sling, and most of the time you'll never know about the con artist until it's too late. I know it comes with the territory. But there's nothing that says you can't use all your available information to protect yourself. And one of the best methods of self-protection is to be on friendly terms with your competition and compare notes.

I'm not advocating telling your competition everything you're doing. I'm talking about the case where it's common knowledge that you and your competition are attempting to sell a prospect, on top of which you start becoming slightly uneasy with the way things are going. You begin to smell a rat; it may be little things that are difficult to put your finger on, or it may be something like the prospect showing you a copy of another salesperson's proposal and sketch. If he does it with your competitor's proposal, then he'll do the same with yours. When such suspicious events take place, you may decide to call your competition and compare notes. A friendly phone call may save both of you some grief.

Don't spend your time worrying about the competition; spend it thinking about how you're going to sell the job. When you talk to a construction salesperson and all he does is gripe about the competition and how they always have a lower price, then I'm willing to bet he's not much of a salesperson.

Price

I waited until now to discuss price. It's too easy to become dependent on price. When you do, you stop selling, and you're back right where you started—on the bid market.

I'm the first to say the price is extremely important; and no matter how super a salesperson you are, you still won't be able to sell any price you want to put on the job. If you are mainly interested in being low, then you're not going to be much of a construction salesperson. An order taker, yes, but a salesperson no!

Out of all the jobs I've sold, I would say that 75% have not been at the lowest price. Nothing else means as much to me in selling construction as hearing from the prospect, "You were a little bit higher than the other person, but I know I'm going to get a better job and that's important to me." It's happened a number of times. When an owner tells me that, I make sure he has service like he never thought possible.

When the selling process deteriorates into a price war, I

try to stress the important things that come with my higher price. I offer to make my price more competitive, but I tell the prospect in plain language he's not going to have this or that service. He's going to get exactly what he pays for, nothing else. Since Mr. X is a businessperson, this statement will make him think. There will be times when price is the only consideration; and if the job is worth it in your estimation, then go after it. Don't get angry when the prospect insists on making a decision on price alone—it's his money.

It's important to keep in mind that the prospective projects are all different in some ways. Each one will have to be handled on an individual basis, and the construction salesperson will have to be flexible in his approach. The best guideline is to do what has to be done to get the job, if you want it. So don't discount price altogether, but don't make price the only item you build your presentation around. The truly good salesperson talks about price last, after he has covered all the things the price includes.

I like to bring into my proposal presentation a thought-provoking idea, and I bring it up before we get to the bottom line: *cheap and economical do not mean the same thing.* What I'm trying to do is have the prospect think about getting more for his money. I impress on him that I'm thinking economy, and not the lowest price. I want that prospect to understand that this was the guideline I used when I put together his proposal. After explaining this, I then say that if Mr. X is really thinking about a cheap building, then I'll be glad to redo the proposal with that fact in mind. Nine times out of ten the prospect tells me he wants an economical job, not a stripped-down one. I've found this to be a very good tactic; use it and I'm sure you'll see what I mean.

I think the best way to sum up this section on selling is to say that *service* is what you're offering the prospect. You set out at the very beginning to impress on Mr. X that you want the opportunity to work for him. Service will count because you shouldn't be offering the low price but a fair price and service. When all else is equal, chances are the prospect will decide who does the job by the service he can expect from the contractor. When you get right down to it, service is all you really have to sell—so stress it, push it, and try your best to make the prospect understand exactly what it is you can do for him.

The next point I want to emphasize is to *follow up.* Contractors have a bad reputation when it comes to getting things done. We are always late with the work and hard to talk to when Mr. X wants to know why we are late. Nothing ever seems to be done when the contractor says it will be. Many people do think this way about contractors, so when you follow up and do what you said you were going to do when you said you would, the prospect has to be pleasantly surprised. Following up will earn you more points with the prospect than anything else you can do.

I'm talking from personal experience. I can show four buildings that I sold where the owners mentioned to me that I was the only contractor who followed up. The others didn't bother to follow up after their initial interviews, or didn't even return the owner's phone call. It's easy to see where the public gets its opinion about contractors. Granted that such contractors are probably too busy to get involved. They should tell the prospect so, not ignore him. Follow up—it sells you and the job.

When the construction salesperson enters the selling phase, it will not be a step-by-step process as I have detailed it in this chapter. Please note that you can't sell construction by the numbers. It's necessary to be familiar with all the parts that make up selling. Then you use what will obtain results at the time you feel it is right.

The salesperson has to be flexible. When you are talking to a prospect, financing may come up before you have a chance to talk about your company references. The prospect may want a ballpark figure immediately. Be prepared for anything, and let nothing upset you. The only way that prospect will know you're a pro is if you act like one.

CHAPTER 7

Contractual Procedures

LETTER OF INTENT

After working as hard as possible, you hit pay dirt when the prospect agrees to contract with you. The question you ask yourself at this point is: Just how do I go about the necessary paperwork? Verbal agreements are fine in some cases but not in construction sales; it's good business sense to have everything on paper and signed by all parties.

The signing-up paperwork is somewhat different from, and much more involved than, the low-bid contract. The reason is that for much of the time the salesperson is working on the front end. You have a certain amount of work to do before you can determine the actual contract price. At the same time, you don't want to be out on a limb and spend money and time without something in writing. The answer for this interim period is a contract to contract, in other words, a letter of intent. (See sample on page 146.)

The letter of intent is exactly what the name implies; the prospect intends to sign a construction contract after certain conditions are met by the construction company. It gives you written protection so that time and money can be spent. You

XYZ Food Co.
Address

XYZ Construction Co.
Address

Re: Letter of Intent
for Construction
of a New Facility

Dear Mr. XYZ Company Official:

It is my intention to sign a contract with your company. It is understood by both parties this letter authorizes you to proceed with plans and specifications to determine the exact cost of the project.

It is also understood the not-to-exceed budget price is $325,000.00 and the contract price will not exceed this amount unless I make substantial changes to the plans.

Preliminary plans will be approved by me before final drawings are made, and I reserve the right to make whatever changes I deem necessary.

Sincerely,

Mr. X

will truly begin to spend money. Your architect will have to be called in as well as surveyors and site engineers. Often if a preengineered metal building is involved and time is a factor, the building may be placed on order. Thus, it's absolutely necessary that you're protected by something in writing.

The sample letter of intent is the type I prefer to use. If you'll notice the letter is written to me by Mr. X, it is on his letterhead and outlines his prerogatives as the owner. Each party knows where he stands and what is expected of him.

In my opinion, the letter carries more weight when it is on Mr. X's company letterhead. There can be no doubt he means what he's saying when he takes the trouble to write a letter instead of signing his name to the contractor's form. I know this is a small point, but it may help; and I know it will never hurt your case if a hassle develops and a third party has to be called in.

Every time I've used a letter of intent, the prospect has asked that I compose the contents. After he checks it over and finds it suitable, the letter is typed and signed by Mr. X. I always take the original for my file. The prospect is aware of the letter of intent before we get to that stage with his project because during the selling process I've explained how we handle the paperwork.

The construction salesperson will encounter all kinds of conditions that will have to be put into the letter of intent. Do anything reasonable to sign up the prospect. You have to nail down the job and get it out of the marketplace as quickly as possible. I've seen very few reasonable conditions that couldn't be worked out in some kind of acceptable manner, so put all the conditions you want to in the letter of intent. Notice I said *reasonable*. As you gain experience, you'll come to realize what is reasonable and what is downright foolish. Obtaining financing is a reasonable condition. Obtaining financing at 8% interest is not.

Even if there is more than one condition, use the letter, and don't hesitate to put all the conditions in it as long as you consider them reasonable. The prime thing is to get the prospect to make up his mind and commit to you. Once this is done, the prospect becomes the owner, and the pure selling phase is over. You are now the owner's contractor, and your main job is to get busy with clearing up or helping the owner to clear up the conditions.

The most used condition is financing, followed by rezoning and permits. Of course, the one condition that will always be in the letter of intent is to bring the contract in close to the budget price. Make sure you understand exactly what this condition entails. Know that the prospect is seeking financing at the going rate, not some ridiculous figure that will be almost impossible to find. If rezoning is a condition, then do your homework and find out if you're going through an exercise in futility. Don't agree to a condition you know will likely not be met. Every once in a while you'll work with a prospect who thinks he has nothing to lose by letting you try to put the project together, though he knows he has an insurmountable problem. Watch out for such ringers because they can really cost you. Chances are if you do a good job qualifying the prospect, you'll know about the problems before you come to the sign-up phase. Nothing is guaranteed, though, so keep your wits about you.

The letter of intent is not worth the paper it's written on, I've had people tell me. You may be wondering the same thing about now. In my opinion the letter is worth a great deal; it is a business document signed by the prospect stating what he intends to do if certain conditions are fulfilled. What the opponents object to is that the prospect has an out. If he gets cold feet, he just tells the salesperson he cannot obtain his financing, and the deal is off. They're right about this.

But then what do you do if after you sign a contract it becomes known that the financing has fallen through and the owner can't pay you? What good is your contract at that point? An attorney once told me contracts are only as good as the parties involved. If the prospect isn't going to honor his signature, then all the contracts in town won't do you one bit of good.

I prefer to think that most of the people in business are honest, and when they sign a letter stating their intentions, they expect to abide by it. If for some reason a condition is not met and there's no deal, then all the parties part friends because it was in the original agreement.

In all my experience I have had only three letters of intent that did not go to full contract. On two of them the prospects very much wanted to build. The first couldn't secure financing; the other had zoning problems that everyone thought could be worked out, including me. The last one was with a prospect who decided to back out. In all honesty, I never should have even tried to sell him and sign him up. I was uncomfortable after getting the letter of intent signed. There was nothing much I could put my finger on, just a lot of little things that didn't quite add up. I was relieved when the prospect backed out.

I don't think three busts are too bad when weighted against the many jobs I've successfully built. Letters of intent are similar to competition; you can spend so much time and energy worrying about them that the job doesn't get done. A letter of intent need not be a one-way street for the contractor; he can have some conditions of his own incorporated. For instance, you include the condition that if the project is not built for whatever reason, the prospect has to pay a predetermined price for the plans. This will allow you to recoup what is usually the largest expense.

The letter of intent comes in all sizes and shapes. You can build it right into your proposal, and all the prospect has to do is sign on the dotted line. You can use a type of form letter where you fill in the blanks; although this is handy, I personally don't care for it. I've found each project has its own particular conditions, and it's much easier to compose the letter to meet the situation.

After some exposure you'll know what works best for you. One good rule of thumb: If the time starts to stretch out after you have presented your proposal, and it's difficult to have a letter of intent signed, then watch out. Every time this happens, the job falls through! The process should not drag. You may not get the job—it may go to a competitor; but don't let the selling drag out. Selling is a percentage game, and you have nothing working in your favor when it drags. Drop it and work on something else.

Plans and specifications will be the first order of the day after you obtain your letter. I'm not going to go into how you get them; suffice it to say that you operate in the construction industry so it should be no problem to have the plans prepared. We are concerned about how the owner (Mr. X is no longer a prospect, but an owner) and the drawings fit together. First of all, there will probably be a time lag while the plans are being prepared. I can only speak from personal experience about how long this lag takes, but it's usually two to three weeks, depending on the complexity of the project. I have had plans in as little as one week for a simple box warehouse.

One word of caution, though: Don't let it take more than three weeks to a month at the longest. I've noticed that the momentum starts to lag after about four weeks. This shouldn't affect the contract, but it could give the owner ideas about maybe delaying the actual construction for some reason. Don't forget you're dealing with businesspeople who won't see things the same way that you do.

I use plans mostly to keep the enthusiasm alive during this phase. I always have the architect prepare a preliminary floor plan to a 1/4-inch scale. Sometimes I'll include an elevation that shows storefront details if I feel it's necessary. I have these drawn as quickly as possible, and I'm back in to see the owner in about a week and a half. During that 10-day period I will have seen Mr. X at least once and maybe twice. There is always something to check with him about.

Be prepared for a very pleasant reception by the owner when you walk in with plans rolled up under your arm. This is the only time I've seen retail merchants tell someone else to take over and escort me to their private office while customers were waiting. Nothing else will turn an owner on like his first viewing of the plans. I think the plans themselves are responsible for this attitude. When the owner finally sees the first tangible proof of the reality of the project, he's like a kid on Christmas morning.

Once I had an owner bend over his desk and with his arm push everything to one side when I presented his plans. Most of the paper and other items ended up on the floor. When I left an hour later, he was just sitting and looking at the blueprints. I hope you enjoy as much as I do this part of construction selling.

The preliminary plans are for the owner to check, and hopefully he will make what few changes are needed. While this is being done by the owner, the site plan work is going ahead; and by the time the owner accepts the building plans, the architect should be ready to proceed.

I use the period of waiting for the final plans to have the owner mark up a floor plan for the phone company and have him confer with equipment suppliers if they have work to be done during the construction phase. For example, in the case of a retail tire facility with car lifts, preliminary work has to be done along with the slab. The owner should be involved with all such situations. It keeps him in the picture. He knows what's happening, and the momentum is steady.

When the final plans are completed, sit down with the owner and go over them in detail. Chances are, he won't know 50% of what he's looking at. But, nevertheless, emphasize the areas that may cause problems and hope he understands. Many times he thinks he understands but doesn't have a clear picture. Those blueprints to you are crystal clear, and construction people sometimes think everyone can read plans as well as they can. Most of the time Mr. X would not even know the bottom of the page if it weren't for the title block, so practice patience. You may find a method I use helpful. When room sizes are discussed, I try to give the owner a visual aid. For example, if his new office is 12′ × 14′, I look around to find some way to show him the size. I may tell him the dimension of the room we are using as a comparison. Size should be gone over with the first preliminary drawing and again with the final one, in hopes the owner has at least some idea of what he's getting.

Many times your best will not be good enough. You've done everything in your power to make him understand the showroom ceiling will be 9 feet high, but when he sees the grid being installed, he'll tell you it's too low; he thought it was higher. You can refer back to the plans as much as you want to, and he'll always say the same thing: "I didn't understand 9 feet was so low; I can't read blueprints. You're the contractor; you should have told me 9 feet was so low." Be prepared for it, and if possible, put a little money in the job to cover this type of problem.

After going over the finished drawings, both you and the owner sign two sets of plans. You take one set and Mr. X takes the other. I always file my set and make it a point not to let it out of my possession. This signed set of plans never gets to the job site; it stays in my office. If it ever happens that the owner tries to put one over on me, then I can haul out the plans and show Mr. X, or, more importantly, a third party, just what the owner has agreed to. I've had to do this only a couple of times, and in both cases everything blew over after the owner saw what he had agreed to.

I said earlier that you may want to have a condition in the letter of intent that reimburses you for the cost of the plans if the job doesn't go to contract. In special cases I can see where this might help make up the prospect's mind to sign. Maybe it's some complicated type of construction that's very difficult to price until the plans are drawn. Therefore, the prospect won't have a handle on price, so he has to take this route. The problem is you may find yourself providing a plans service, and there's no way to get back what it's worth for you to solve the initial problems that precede the letter of intent.

I really don't like any agreement where the prospect can buy the plans and go his merry way. I stay away from this and do it only when I think the job is worth the calculated risk. I'm the first to say this is another of those gray areas the construction salesperson encounters. The only guideline I can offer is to use common sense and know who you're dealing with. If you do enough construction selling, you will be taken advantage of every once in a while. It comes with the territory.

I personally know of some contractors who will prepare plans and tell the prospect to use them to take other prices. If their price is not low, then they are paid for the plans and the owner contracts with the other person. My answer is that I wouldn't do it under any conditions, period. You're right back on the bid market—and providing the plans as well.

The prospect can sometimes find more objections to signing than you can imagine. Most of the time this comes from a good case of being a little overcautious. You have to sign the prospect up while giving him an out. He will not be completely sure until he has the bottom-line contract price. Recognize this fact about the person you're dealing with, accept it as part of the program, and go ahead and work around it. When Mr. X knows the final contract price, then he won't hesitate to commit completely. Don't try to force something on him that he's not comfortable with. If you do, you can blow the whole deal.

There is one other way to sign up the prospect besides the letter of intent—I call it a simple letter contract. It's very practical for simpler jobs where you can nail down cost with your sketch. I use this mostly with metal buildings when there is very little subcontract work to be done—just slab, building, and erection. It's very easy to give the prospect a firm price under these conditions. The prospect can see at once exactly what he's buying, and plans are not necessary to close the deal. Mr. X signs on the dotted line, and you're in business.

I've found from experience that the prospect is very comfortable with this type of special contract. He doesn't have to have his lawyer check over the fine print because there is none. It's clear-cut, and he knows where he stands. I even include a payment schedule so that all parties know when the money has to change hands. Don't try to cram too much into a letter contract. As soon as it loses its simplicity, its effectiveness is also lost.

It is possible to sign up the prospect with several other contract variations that allow the prospect to get the project moving without a concrete price being determined. There can be dozens of reasons for using one of these methods.

The first is the age-old standby, cost-plus. There's no need for me to explain how it works—you're in the construction business. If you can get a cost-plus contract, then good for you. It's an ideal way to do a design-build contract.

The other two types of sign-up contracts are actually similar. There's the contract not to exceed an established figure. I've used this type only a couple of times. The problem is that to protect yourself, you have to jack up the price. This higher price then may become an obstacle, so it requires a special situation. For example, once I used this type of contract because the prospect was under the gun with a deadline, and he dealt in perishable goods. For this reason he couldn't rent suitable space to move into on short notice; he had to have the new place. So in order to crank up the project at once, we agreed on this particular form of sign-up contract. I remember we went to work on the site with a line drawing of the building layout; by the time the plans were drawn and the permits issued, we had everything ready to pour the foundation. I might add that the city inspectors were very helpful and understanding. You may have a chance to use this type, but I'm sure you won't do it very often.

The next method is a variation of what I have just discussed: contract not to exceed established price, with savings shared. I confess I haven't done a project using this idea, but I do think this type of contract has merit. It should make the prospect comfortable to know that his contractor has an incentive to try to save money. The main purpose of this contract is speed. This is the only reason, in my opinion, a prospect will move ahead without knowing the firm price of his facility.

The whole idea in any case is to get the prospect to commit to you and allow you to put the project in motion. Use whatever method will obtain results. Familiarize yourself with all the various ways to sign up the prospect, and don't hesitate to use something new you feel will do the job. Each prospect is different; you are dealing with individuals in a one-on-one relationship. Approach each one with his requirements and problems in mind.

SPECIAL LETTER CONTRACTS AND FORMAL CONTRACTS

The prospect has been sold and signed. You've solved all the problems. The first price has been put together. It's now time to have the owner put his name on an iron-clad contract so you can start work. What kind of contract do you have in mind? What kind of question is that, you ask? A contract is a contract. What's the big deal?

The big deal is that there is more than one type of contract. I'm not really concerned about the legalities of the contract, but how its form is accepted by the owner. From experience I know the simpler the contract, the quicker it's signed. This is most important because you don't have a job until the contract is executed. The longer this takes and the more new parties brought into the picture at this time, the more chances there are for delays and problems.

I imagine most of your contract experience has been with long, very detailed documents following the American Institute of Architects (AIA) or Associated General Contractors (AGC) format. The contracts may have been the type furnished to general contractors, or the contracts may have been prepared by the owner's lawyer in conjunction with the architect. Maybe your attorney drew up the papers. In any case, legal jargon prevails, and the contract is very likely to be more difficult to read than an insurance policy.

Another situation you may or may not have encountered when using some of the standard contracts in the construction industry is having to mark out whole sections that are not applicable to a particular situation. The owner can become uncomfortable when he sees a large legal document or when he looks at a contract with sections marked out. Mr. X is down to the wire now and he wants to know that everything is exactly as it should be. When the owner isn't sure, he turns to his lawyer and asks that the contract be checked over. The problem here is time; an attorney is not going to peruse the contract lightly. He's going over it with a fine-tooth comb because he's paid to protect his client. This takes time, and everything stops until lawyer and client are satisfied with the contract.

If you are seriously going to pursue selling construction and you do not use a shortened, easy-to-understand contract, I recommend you look into the matter. From experience I've found out it does help move the project along.

I personally prefer the letter contract. (Note sample on page 160.) It's short and to-the-point and can be easily expanded to fit the job. A simple job consisting of a metal building on a slab for industrial storage takes up a page and a half. On the other hand, a more complex project may take four or five pages detailing the work with spec sheets attached. The important fact is that it all takes place in the letter contract form, which keep things easy to understand. This means the selling process goes smoother.

As I pointed out, the letter contract on the simpler, easy-to-price jobs can eliminate the need for a letter of intent. Added to this is an interesting fact concerning repeat work. Hardly ever will a letter of intent be needed when you are asked back to do work. The fact that you have established a very good relationship with Mr. X of which mutual trust is a large part will preclude the letter of intent. It's quite pleasant to get a call to stop by to work up the details, and then go from there to the contract. So my advice about repeat work is not to use the letter of intent except in very special cases.

There is one negative condition the owner always seems to get around to when the contract is to be signed: a penalty clause for late completion. It comes up so often I now bring the subject up for discussion before the owner does. I've never had to construct a design-build project where I worked under a penalty clause. I stress to the owner that the design-build job has the advantage of constant attention; if there are any delays, they will be due to matters beyond my control, such as weather or strikes. In addition, I always tell the owner I do construction for a living; if I don't perform as I say I will, then the job starts costing me money. So it's to my personal benefit that I finish the project as quickly as possible.

Mr. X, being a businessperson, will usually accept these explanations; but every now and then you'll get an owner who'll bear down on the penalty clause. Here I use a tactic that's never failed. I tell him that since he insists on a penalty clause on completion date, and is a fairminded businessperson, I'm sure he'll be willing to agree to a bonus for early completion if I'm willing to agree to a penalty for late completion. No one has taken me up on my offer.

I'm not trying to tell you how to run your company business, but I am telling you ideas that I've found to be very helpful in closing the deal. I've talked with people who manage salespeople, and they all tell me the same thing: one of the hardest things to teach salespeople is the art of closing the deal. It's amazing the number of good salespeople who have trouble getting the prospect's name on the contract. I don't think the word *good* should be used in this case, for if they can't close, then they certainly aren't good salespeople. When sales managers discuss personnel, the bottom-line question is: Can he close? Only closers get the fame and fortune; all others should try some other occupation—there's no way they can be considered salespeople.

The worst mistake a construction salesperson can make is to become complacent after he has the letter of intent. He must keep right on pushing ahead because he has not closed the sale until a contract is executed. With this goal in mind, I prefer to use the simplest contract possible that will allow me to close the deal as quickly as possible.

I have several tactics I use when presenting the contract. One is taking the approach that even when the owner obviously wants to get the project moving, he still has a tiny cloud of doubt. Keeping this in mind, I don't just hand the papers to him and wait for him to sign. As far as appearances are concerned, my main thrust is not getting the contract signed. I take the attitude that the contract is as good as signed. I bring up subjects that will require the owner's help or a decision very soon. If a metal building is involved and the color has not been chosen, I ask the owner if he can have the color by a certain time because I plan to order the building just as soon as I get back to my office. Notice I don't say just as soon as the contract is signed. The impression I want to make is that it is a foregone conclusion the contract will be signed. I am mainly interested in taking care of the loose ends.

Another good topic to bring up for discussion is the coordination between general contractor and people installing special equipment who are working directly for the owner. A good example would be a shop you plan to build that will have a bridge crane. The owner is dealing direct on the crane, but you have to provide special supports for the rails. Then when all else fails, and I really have nothing to ask the owner, I fall back on inflation. I tell the owner that his timing is perfect; with what inflation is doing to construction costs, he would have to pay a lot more for his facility had he decided to hold off. I use this only when I have nothing else to bring up, though I've found the owner will sometimes mention it while we are talking about details.

Don't be bashful about asking for the contract. A good closer will ask for the signature at the right time. It's easy to work it right in while going over details. I say something like, "Mr. X, I brought only one copy for you to keep. If you need any more, I'll drop them off tomorrow." (Notice I did not say I would fax it.) While talking about the contract, I suggest that we go ahead and take care of the paperwork while we are on the subject. I learned from another salesperson that it's a good idea not to have signed yourself; doing so at the right moment will start the ball rolling, so to speak. I've used this suggestion successfully a number of times. Consider trying it.

Some of you may think all this to-do about how to handle the contract signing is a lot of nonsense. My answer is the same thing I've repeated so many times before: Selling construction is strictly percentages. You're always trying to have the percentages work for you. Consider this: In 90% of your contract closings, you are probably right; it's not necessary to go into such elaborate selling maneuvers. But for the other 10%, it very well could make the difference. The hitch is: Will you be able to tell which owner is in the 90% category and which is in the 10%? I can't! So I assume everyone is in the 10%. I believe my approach will never cost you a contract. It can do nothing but help.

No matter what form of contract you use, they all say the same thing: what you will be paid for performing a certain job. And this leads us into another item that should accompany the contract: payment schedule. The payment schedule is important in a design-build project because the general contractor has already spent money to arrive at the contract stage. Also, there will be anticipated expenditures to start up and man the job site. If a preengineered metal building is involved, a deposit may have to be sent to the manufacturer before he will fabricate the building. To cover these costs, I always ask for and get a deposit when the contract is signed or very soon afterward.

Don't just spring this deposit on the owner. After the letter of intent and before the contract is ready, I discuss how the money is to be handled and exactly what the owner can expect. My rule of thumb for the deposit is 5% of the contract amount. This 5% is flexible, though; sometimes I'll decrease it, depending on the project and the owner.

I take pains to explain to the owner exactly what the deposit is for—to pay the design professionals, and to pay my company for time and work done to date. The owner is a businessperson, and when you explain things in terms he can relate to, there's no problem. He's in business for the same reason you are—to make money.

RETAINAGE AND BONDS

I know you are familiar with retainages. They're something we have to put up with in the construction industry. You'll be faced with retainages in design-build as well as bid work. It's brought up by the owner, and if he insists, I put a 10% retainage clause to be applied to each invoice. I only do this now when the owner prefers it; usually when the subject comes up, I try to persuade the owner that it's not necessary on a design-build contract because he knows who he is doing business with. It is really used for owner protection on the bid market, where the contractor is probably a complete stranger to the owner. Frankly, most of the time the owner accepts this explanation, and there is no retainage.

However, there is one thing I make clear when I do have to work with retainage being withheld. When the job is complete and the owner moves in, the full amount is due. No retainage will be owing when Mr. X enjoys the use of his new facility. Callbacks will be handled on a warranty basis, not by holding my money as a club. Also, the question of a bond may come up, if needed. Then I tell the owner the cost of the bond will be added to the contract as an extra.

CHANGES

This is a good place to discuss a peculiar situation that you'll run into if you do enough construction selling: that of the owner who thinks after the contract is signed and the job is coming out of the ground, he can start asking for little favors that won't cost him anything. Maybe you've already encountered this type of owner. In defense of some of the owners, they really have no idea what it costs to do something and think a favor or a small change won't make any difference. There's only a few bucks involved. Then you have the sharpie who's going to try to squeeze something out of you for nothing.

In the first case, if I can do what the owner asks and it involves only a few dollars, I do it. But if it's going to run into some money, I make it a point to explain, diplomatically I hope, that the change will cost some dollar amount. Then I tell him why. It could be very little material cost and high labor, or any number of variations; but the point is that I explain to the owner my situation.

I deal with the squeezer in a more blunt fashion: I put everything in relation to his business. He might be a car dealer, in which case I ask him how many extras he gives away when he sells a car. In other words, ask the owner what he gives away and he gets the picture very quickly.

CHECKING OWNER FINANCING

Most of the owners will have a construction loan, so it becomes necessary to provide invoices at the end of the month for the owner to turn into the bank before he makes his draw. Therefore, I make a point of explaining exactly what Mr. X can expect at the end of the month. Check with the owner's bank before the first draw is due so that you will submit an invoice that includes everything the bank needs. You certainly don't want any delays in being paid!

The payment schedule can be separate and attached to the contract, or it can be incorporated into the contract itself. The letter contract can have the payment schedule included as well. (See example on page 161.) It's important that all parties know how the money will be handled. Always ask for and get your front-end deposit. It's a comforting feeling to be operating on the owner's money right off the bat.

I can generally put up with some headaches from the owner if I know the money is there and waiting. The way I do this is to ask for a copy of the mortgage commitment, and confirmation from the bank making the construction loan that everything is in order. I like to do so diplomatically because I'm faced with the same problem I have when presenting the contract to the owner: 90% are going to sign; the remaining 10% may have some reservations.

The problem with getting into the owner's finances is that some businesspeople are touchy about their word and their credit. Every now and then you'll encounter someone who becomes annoyed when you doubt his word by asking for proof. Most of the owners accept providing proof as the way business is done, especially when dealing in the sums of money involved. Again the problem is who's touchy and who's not. This becomes important because the situation usually comes up before the contract is signed.

I've developed a tactic that works: I become an agent for a third party who needs the information. The first thing I do is take the offensive. I point out that I have to have a copy of the mortgage commitment. Then I explain that my bonding company would like to have the information, or I might use my suppliers. I explain that because of the large amounts of material I'll be buying from them, they want to know the bills can be paid if I, the contractor, drop out of the picture for any reason.

Now Mr. X may get a little annoyed, but not at you—rather at a third party who is just a name. You come across as a nice guy trying to do your job, and these other people have put you in the middle. Mr. X does as asked more to help you than for any other reason.

This chapter is for the benefit and protection of the contractor. A signed contract is the target you were shooting for way back when you started chasing down leads. It's the bottom line, so use my suggestions if you feel they will help. Along with the contract, make sure the owner knows how you expect to be paid and protect yourself by knowing the owner has the money to pay you. Assume nothing; no one else will look after your interests the way you will yourself.

Company Letterhead

Date

Mr. X
ABC Distributing Company
Address

Re: Construction of new 6,000-square-foot metal building

This document is a contractual agreement between ABC Distributing Company and XYZ Construction Company to construct the above-noted building.

Details are as follows:

Metal Building:

1. Furnish and erect one (brand name) preengineered building 60′ × 100′ × 16′. Roof slope to be 1:12.
2. One (1) 3070 metal walk door with lockset.
3. Gutters and downspouts.
4. Two (2) metal overhead doors, manually operated 10′ wide, 12′ high.
5. 3″ white vinyl-backed insulation in roof and walls.
6. One (1) 3070 glass storefront door.
7. One (1) 6′ × 6′ and four (4) 2′ × 5′ insulated windows.
8. Three (3) future overhead door locations.
9. Eight (8) 3″ × 10″ skylights.
10. 4′ roof overhang in front with soffit.

<u>Concrete Work</u>

11. Foundation will be steel reinforced 3000# concrete; slab to be 5″, 3000# concrete with 6″ × 6″ × 10 wire with 6 mil poly vapor barrier; joints to be cut; all in accordance with plans by a registered architect.

<u>Contract Price</u> $90,000

12. Payment Schedule $ 4,500.00 Signing contract; $42,000.00 7 days after building delivery

Monthly billing for all other work. Construction time approximately 90 days.

<u>Items Not in Contract</u>

13. No interior work in building (electrical, HVAC, plumbing, etc.).
14. Cost of permits, city fees, and site plan.

<u>General Information</u>

15. All proper insurance as well as builder's risk in force during construction period. Certificate to be supplied.
16. Plans by a licensed architect.
17. Standard construction industry one-year warranty on all materials and labor.
18. Contractor will not be held responsible for any delays due to weather or areas beyond his control.
19. Building to meet all building code requirements.

ABC DISTRIBUTING COMPANY	XYZ CONSTRUCTION CO.
______________________	______________________
Mr. X	Company Official

Let me sum up this chapter on the paperwork by stressing the fact that I am sure you are familiar, and have plenty of experience with construction forms and contracts. This means you will have no trouble coming up with what will do the job to get the project signed up.

The main trust of this chapter is to combine selling with signing up. I strongly feel it is just as important to be selling you and your company here as in any other area of design-build marketing.

So regardless of what particular form of paperwork you choose to get the project moving, I suggest you keep the selling recommendations in mind.

CHAPTER 8

BUILDING PHASE COMMUNICATIONS

With the signing of the contract you've accomplished what you set out to do—sell a design-build project. The active aggressive selling is ended; all that's left is actually building the facility. Right now if you're like 99% of the construction companies, you file away the contract, assign a superintendent, and move on to something else. And if you do, I'll guarantee that at some point during the job you'll have trouble with the owner.

Contractors have a difficult time realizing the owner in a design-build project has to have his hand held from start to finish—I mean right up to the grand opening. I think maybe the standard attitude of the contractor comes from the manner in which he has conducted his bid business. The owner is really a stranger kept at arm's length with the plans, specifications, and contract governing every move. It's natural for the contractor to operate this way; it's the way it's always been done. From a sales standpoint, however, it's absolutely the worse way to handle a design-build owner.

On the other hand, the owner of the construction company has his problems too. He can't drop everything to look after one owner who might get his feelings hurt. Granted he can't stop everything, but it's still possible to give a little extra attention if the effort is made.

START-UP

Within a week after the contract signing and preferably within just a couple of days, there should be some sign of activity on the site. The best thing to use is a job sign, and a trailer if one is required for the project. There can be real trouble if nothing seems to happen on the site; the owner will get upset, which many times will set the tone for the entire job. It's important for the owner to see attention given to the job.

During this period check with the owner to see if he plans a groundbreaking ceremony. If he does, then offer your help, of course; and, more important, work out the timing so the job signs are up and the top person in your construction company is there to get his picture in the paper. This is great free publicity; take full advantage of the opportunity. Don't hesitate to bring up the topic with the owner. He just might decide to go ahead with your help, and the pictures in the paper will be money in your pocket.

Along with these visible signs that the project has started, make every effort to get the project out of the ground. Any good construction person knows it's to his benefit to get going, but things do have a tendency to drag sometimes. If the problem is manpower, then make sure the design-build job takes preference over a bid job, if possible. Treating the bid owner in the regular manner won't change things, but treating the design-build owner in the regular way could cause problems.

WORK SCHEDULE INFORMATION

This project is the biggest thing in the owner's life right now. He expects to be kept up-to-date on progress. Often that's vitally important to the owner because he has to coordinate the moving of his business; the buying of new stock, the delivery of new equipment, and even being able to terminate his lease all make it imperative that the owner have a good idea of the construction schedule.

I've learned that my presenting the owner with a truthful schedule is something he very much needs and appreciates. The standard bar chart used by the construction industry does the job nicely. It's easy to read and gives the owner a feel for the whole job at a glance. Just about anybody can read the bar chart, but there will be the isolated case where you may decide simply to list the schedule with the anticipated dates.

Do not use the critical path schedule unless you know the owner is at home with this type of graph. I tried this schedule once. The owner was too embarrassed to admit it was still like a foreign language after I had explained it to him, so he went through the entire job not knowing what was going on except for what I told him. To you and me it's clear as a bell, but to that owner it was a jumble of lines that meant nothing. Just like the blueprints, I might add.

Provide Mr. X with a work schedule he can easily understand; don't include very much detail. Just cover the main areas.

ACQUAINTING PEOPLE ON SITE WITH THE OWNER

There's no way you're going to keep the owner away from the site during construction. Don't even try. Instead prepare for his being there. Always let your superintendent or foreman know if possible that you and the owner or the owner by himself will be coming to the site. Make sure you introduce the owner to your key people on the site. Make sure the construction people understand Mr. X is the owner and that when they see him on the job, they are to make it a point to speak and try to be helpful. Under no condition is the owner to be ignored. It's the responsibility of the person in charge to take care of the owner when he arrives.

Let me digress for a moment with some friendly advice for the construction salesperson who will be working for a construction company. First realize that the bulk of the hired hands will feel just a little bit of resentment toward the salesperson. There is no way you can convince them you work just as hard as they do. They see you as a good-time Joe riding around with owners showing them jobs and then maybe playing golf on the nice afternoons. You don't get your hands dirty, or freeze in the winter and bake in the summer. These fellows know beyond a shadow of a doubt that you make as much money as the owner of the company.

With the workers having this opinion of you, take it nice and easy when you visit the site with or without the owner.

Treat everyone as an equal; get to know the foreman if you don't already. Don't tell the construction people how to do their job or act as a spy for the owner of the construction company. If something is obviously wrong, question it; but don't go running to the office. Let the foreman handle it; that's his job. I always made it a point to ask the person in charge on the job site if there are any questions I can help with. Invariably on a design-build project changes have been made that create questions for the people driving the nails. In like manner, when a change is made during construction, I visit the site and, over coffee, update the superintendent. This makes his job easier and cuts down on costly mistakes. At the same time it shows the nuts-and-bolts crew you're actually involved and are working just as hard as they are at keeping the job running smoothly.

ACQUAINTING THE OWNER WITH CONSTRUCTION PROBLEMS

Explain to the owner that everything is not going to be rosy all the time. It's important that the owner be made aware of the basic problems that always seem to slow a project down. The only yardstick the owner has is his own business, and chances are that is an inside operation. The weather does not shut him down; and when he has a problem with his goods, he takes another item off the shelf. He is used to solving his business problems by taking immediate action.

I explain to Mr. X by relating my problems to his business. I ask what would happen if an inspector came in and, because of some fine point in the interpretation of the code for selling tires, made him shut his door until the matter was corrected. How would he like the idea that every time it rained he had to stop business, his production came to a halt, but he had to continue to pay some of the help? After a few minutes of this the owner starts to have a better appreciation of what his contractor faces. It really is another world!

During the construction phase keep the owner up-to-date on the bad news if it affects the completion date. Don't let him find out from someone on the site that the steel will be two weeks late. You'd better know yourself and then go to Mr. X and tell him, and just as important, tell him why.

At times a picture will be worth a thousand words. Take the owner to the site and show him the situation. I remember a few years back it seemed to rain every day for three months. We were getting further and further behind, and there was no way to catch up and make the projected completion date on a particular job. The owner was getting upset with our lack of progress. He simply could not see why, when it stopped raining, we couldn't get anything done.

I finally decided to show him exactly what we were faced with. I took him to the site where the half-finished building stood surrounded by a sea of mud. Sitting in the mud with the bottom half of the tracks out of sight was a bulldozer. I told the owner that we had tried to use the bulldozer to reach the building that morning, but he could see how far we went. I didn't hear another word from him about our slow progress.

KEEPING THE OWNER IN THE PICTURE

As stated earlier, the owner must not get the impression that suddenly he's no longer important. It'll be up to you to keep Mr. X very much in the picture.

At the end of the month when the invoice is prepared, don't just mail or fax it to Mr. X. Instead, call and ask if you can pick up Mr. X for a job visit. Tell him you have the monthly invoice and you want to check it out with him. Now the invoice should list the work that has been completed or percent of completion if it is still in progress. The idea is to show the owner what the invoice covers. This gives Mr. X a good indication of just what progress is being made. Chances are the invoice will end up at the bank, and the bank people will inspect the site for themselves. Nevertheless, your customer is the owner, not the bank, so cater to him.

As the job progresses, small questions will come up that may or may not require a decision from the owner. For example, the swing of an office door is in question because a vent had to be moved; or there is a question about where a vision glass should be located in the shipping foreman's office. These are the little questions that put the owner on center stage and keep him feeling that he's very much a part of the project. Even if you already know the answer, file the question away and use it when it's time to involve the owner.

Whenever possible, have the owner visit the site to make the decision, but if that is not always convenient, go to Mr. X's place of business with your plans. The owner should see you at least once a week. Try not to make up trivial little things to present to the owner. Use only legitimate questions, as the true professional should.

When it's time for interior selections to be made, such as paneling for the walls, paint colors, and floor coverings, you have a very good reason to consult the owner. Again, don't just drop off the samples. Go over the choices, and if you know where a certain item has been used, offer to show it to Mr. X. Stay involved and be helpful at this stage. One word of advice, though: do this well in advance of when it is needed or you may be in a pinch for time, and the owner could end up slowing the job down until final selections are made.

Don't worry too much about what you'll use for reasons to call on the owner and keep him in the picture. With a design-build project, it's amazing just how many situations come up that have to be checked out with him. The point to remember is that the owner has to be kept in the picture.

OWNER PROBLEMS DURING CONSTRUCTION

As much as you and I would like to think that this subject is unnecessary, unfortunately that just isn't so! Take note, and believe me, you will have problems with the owner. The very best you can do is anticipate them and be prepared; hopefully staying on top of the job will allow you to catch a little problem before it blossoms into a monster. Below are the major problems with the owner that I have encountered, along with methods to handle the situations.

The Plans Don't Agree with the Owner

This problem always seems to come up, and late in the job when Mr. X can actually see exactly what he's getting. The basic problem is the inability of the owner to read blueprints. It's nobody's fault; it comes with the territory, and the contractor has to accept it.

I touched on this problem earlier in the book, but I feel more detail is needed. The first defense you can use is to keep drumming into Mr. X's head what the plans call for. The second and third defenses are what I mainly want to cover. The second defense is to be astute enough when going over the plans with the owner to recognize when he's really not getting the drift of what you're explaining. He might say he does, but you know differently. This takes a little experience, but it does come with doing design-build jobs. When you sense this, or know it's a complicated area you're discussing, write it down to follow up. When the area in question is being built and the shape and size can be easily seen, bring in Mr. X and go over the section so that Mr. X can see and understand what he's getting. The trick is to do this soon enough that if there are changes to be made, it won't cost too much.

When the question does pop up, you have to fall back on the plans. They show exactly what you priced and will be the basis for a change order. The owner may grumble some about assuming this and thinking that, but he'll be able to see your position.

Every now and then you'll get an owner who won't be satisfied with the way the plans read. He'll keep coming back to a difficult-to-answer statement: "That's not what I told you I wanted; never mind pointing to the plans. I know what I asked for." What do you do? Fall back to the plans and prepare for a fight? Not if you can possibly avoid it. This leads to our last defense—the slush fund, which is a small amount you sandbagged in the job to cover such an unpleasant situation if, when down to the wire, you felt you had to give. Don't give until you've tried every avenue of compromise, such as offering to make the change for cost, or maybe splitting the cost. In any event, for the sake of good relations, use the slush money. Cheer up; maybe the owner will ask for a legitimate change order and you can get your money back.

A great deal of the way this unhappy situation is handled is up to you and how well you know your owner. I haven't had many owners who really presented a problem, and every time I did what they wanted, it was simply for public relations. Sometimes I gave more than I should, but I was not forced to. There is a difference.

Change Orders

I know you're familiar with how to prepare change orders: in writing, with everything spelled out and signed. But I would like to mention tolerance. Sometimes it's easy to become upset with the owner over a change order. It's a pain in the neck for job progress to be interrupted by a change order. Nevertheless, the design-build owner deserves and should receive cheerful handling of his change orders. Give him the same consideration you gave him before the contract was signed. It will all add up to recommendations in the future. Never forget how important this is.

The Construction Schedule Falls Behind

No matter what the general contractor's best intentions are, the majority of projects seem to fall behind. I firmly think the reason is that time is used as a selling tool. I know I've done it! You don't want the prospect to become discouraged over the excessive time span, so you present the best possible construction time frame and hope you don't stumble. Well, somehow you do stumble, and the job starts running behind. The owner, with all his commitments, suddenly sees a crunch coming down the road and starts raising hell!

I know you've been in this box before, but let me tell you how I handle it. Even when the job is coming out of the ground, I tell the owner every time there's a problem beyond my control that costs the job time. Then Mr. X is aware and can judge the commitments he's made for his openings.

The weather, more often than not, is the main reason the job slows down. I use what I consider the ideal way to handle this with the owner. The idea comes from a company I worked for. The weather report was clipped from the paper and filed every day. At the end of the month a tally was made of the days it rained and of the clear days. The strength of the wind was also noted, since it has a direct effect on metal building erection. These weather notes will give you good backup when meeting with the owner, so take the time to clip the information.

Now the owner has no trouble understanding bad weather as the reason, with one exception. Just as soon as the rain stops, he expects the job site to be covered with workers. One of the hardest points to get across to Mr. X is that the side effects of the rain are just as damaging as the rain itself. This is why I suggest when you explain to the owner some of your problems, you stress the side effects of rain, and how it can cripple a job. Even then you get phone calls from Mr. X demanding to know why there's no work being done, since, after all, it stopped raining 20 minutes ago, my advice is—be nice!

On this subject of falling behind, I've made the assumption that the delays are beyond the contractor's control. If they are within your control and the job is behind because you're not pushing it as you should, then you'll have a real problem with the owner when and if he finds out. This is the one area where the entire relationship you have so painstakingly built up with Mr. X turns into a can of worms. From this point on you're just like any other contractor, and the owner feels justified in doing anything he can to protect himself. Guess what Mr. X uses for protection? If you said money, you're correct. He starts dragging his feet on payments just as the job is dragging.

Finishing the Job and Punch List Performance

In my opinion the last 5% is always the most difficult to get completed. The project is winding down, and hopefully you are involved with new jobs coming out of the ground. Therefore, Mr. X's job is receiving less and less of your attention.

It's extremely important for the contractor to work at staying on top of the finishing-up process. I know personally how hard it is when you have a full load, but it has to be done. I learned the hard way by having a couple of jobs turn sour in the last 5% stage.

Not only do you stay with the project, but give a little more time to the owner. It's a very trying time for him. He's facing the hassle of moving his business, plus having to coordinate with his suppliers. Some personal attention and help with getting the building ready will go a long way when referral times come round.

When the project is very near completion, arrange for a walk-through meeting with the owner, as well as the foreman responsible for seeing that the work is completed. Give yourself plenty of time. These punch-list walk-throughs can run into time because the owner will be enjoying his new building. After the punch list is worked up, get all the work done in a reasonable time frame. Arrange a second meeting to go over the completed punch list. Check the job ahead of time so there won't be any surprises.

Finally Mr. X is in his new building, but the contractor's work is not done yet. Two other matters should be attended to. The first is the owner's grand opening. If there is one, be there. Mr. X is pleased as can be with his new facility and wants very much to show it off to all comers. As close as he's worked with the contractor during the job, he'll feel slighted if the contractor doesn't take the time to attend and help him enjoy the occasion. There's another reason to show up: it's a great place to pick up leads, so mingle with the crowd! There's a good chance some of Mr. X's competitors will stop by to see the new facility, enabling you to do a little self-serving lead gathering.

The second chore left for the contractor is to follow up in about three months with the owner to check on any problems.

Most owners have never known of a contractor to check back. This follow-up call really impresses the owner. I've picked up several good referral leads from an owner while making this type of call. You may wonder why the owner didn't call you about the lead, but probably he just forgot it until he saw you.

Also there will be callbacks right after the owner occupies the building. See that they are taken care of, and then personally check back. It's this kind of service that reputations are built on; it means business in the future.

The contractor who sees the owner through all the trials and tribulations of construction will be making sure that owner becomes a ready source of business for him. The happy, smiling owner will be your best salesperson. This happy result doesn't come free; you have to work for it, but the rewards are well worth it.

CHAPTER 9

THE DESIGN-BUILD SALES DEPARTMENT

This chapter is intended primarily for the construction executive who is interested in forming a sales department. When I mention sales department, it can mean one or more salespeople. If you have or are planning to employ one salesperson, and his responsibilities will be selling, then you have a sales department.

I have no doubt whatsoever that more and more general contractors will be using professional salespeople in the future. As time passes, we will experience many changes throughout the industry, and construction selling will be a part of these changes. I hope my thoughts and the experiences that I relate in this chapter will help you form and operate a successful sales department.

The first thing I should explain is the type of person you'll employ as a salesperson. You'll have someone working for you who will look like he's not. He'll wander in and out of the office on no set schedule. Most of the time you'll have a vague idea about where he is and what he's doing. Some of the time you won't have the foggiest notion of where he is. You may go for a couple of days without seeing him. You call his home in the middle of the morning to leave a message and he answers the phone. All this can be difficult to accept if you are used to people punching a clock. But stop and think for a moment. Where is the one place your salesperson is not going to sell a building? Your office! The office is something he will use to help him. It becomes only a backup operation to his main working area, which is everywhere except the office.

I'll never forget what one of the owners of the firm where I first worked said to me when I started. He told me the one place he'd better not see too much of me was in the office. Neither he nor anyone else in the office had any intention of buying a building from me.

The truly professional salesperson is very independent and wants very little to no supervision. This is even applicable to firms that use a formal sales force, such as the building material manufacturers. They have sales managers and all the necessary chains of command, but their sales manager knows when he has a professional working for him. He doesn't have to read the paperwork to tell him what the person has been doing; he only needs to glance at the order sheet because that's the only yardstick one can use to measure a professional salesperson. Does he get results? That's the bottom line; everything else is just window dressing.

A good one-word description of the construction salesperson is *maverick*. He's his own person, a self-starter who's good at what he does and knows it. The construction salesperson is told what he is expected to accomplish but not how to do it; that's up to him.

Okay, you say, I get the picture of what I can expect in this individual; but how do I fold him into a construction company operation, and, just as important, where do I find such a person? The rest of this chapter will help with these questions.

INTEGRATING THE SALES DEPARTMENT INTO THE OVERALL OPERATION

Your first step will be the physical part of integrating a sales department into your company. The salesperson needs a place where he can take care of paperwork and make phone calls. An office is fine, but if one is not handy and you have a conference room with a phone, there's nothing wrong with using it. Matter of fact, I used a conference room for two years as my office.

When the conference room was in use, I took over any empty office. This worked very well, and the firm didn't have to rent more space, which helped on the overhead.

After giving your salesperson a place to hang his hat, see that the clerical requirements are covered so that necessary office work will be done. With regard to typing, it's imporatnt to give the salesperson some rank. He needs to have some authority so he can have office help on short notice. And I mean short notice! When the prospect tells your salesperson to type up a letter of intent for him to sign, it had better not end up on the bottom of the typing pile. It goes right to the top and is finished promptly.

The salesperson has to be able to have a short turnaround time in the office when he needs material to get a job moving. Many times it's simply a case of asking the typist to please type the sales proposal next. In larger offices where there is an office manager, let the salesperson go through the manager. You have to make clear to the manager, though, that the sales request will be treated as priority when he is asked that it be handled in that manner. Make sure your salesperson understands not to ask for priority treatment unless it's important. Salespeople have a tendency to be impatient when it comes to paperwork, expecting everything to be done for them immediately. There will have to be a little give and take here.

The office help won't understand why this new person has to have a letter typed in the next hour. They are used to having the paperwork move at a set pace. The people doing the work and the people receiving the material all know how the system works. Nothing, they will think, has to be done that quickly; after all, the salesperson has been working on that particular prospect for the past four months, and a couple of days certainly won't matter. Well, it just might matter a great deal, especially if it's toward the end of the week. You will have to impress on the office people just how important some of the paperwork will be.

Even then be prepared to referee every so often; for if your salesperson is good, he'll take matters into his own hands when necessary. For example, I had been working on one prospect for six months and I was determined to sell the job. One Friday morning Mr. X told me to work up a letter of intent and general specifications to go with the drawing I had furnished. I asked if he would be available that afternoon if I could get the papers typed up for his signature. He replied yes, but what was the hurry? I told him I wanted to work on the job over the weekend, which was only partially correct. I also wanted the papers signed before the weekend.

I drove directly back to the office, dashed in, and found no one there except the three people who worked in the office. It was the end of the month and they all were working on invoices to send out. The office manager was a man who was very sympathetic to the plight of the salespeople (there were two of us) except when he was preparing month-end invoices. Then he guarded the typists' time like a mother hen. There I was with a hot prospect ready to sign, and no one to tell the typists to type for me. And their instructions were implicit: type invoices and nothing else.

I took the attitude that the invoices of sold jobs should take a back seat to one that needed selling. I told two of the typists to stop what they were doing, and I put one to work on the letter of intent (the draft of which I had been carrying around in my briefcase for a month) and the other to typing a specifications list. Just as they were finishing the typing, in walked the office manager. He took one look around and did a marvelous job of holding his temper. I snatched my completed papers, thanked the typists, and bailed out. One hour later I was back with everything signed.

I also knew I would be hearing from the office manager, and, sure enough, I'd no sooner walked out of the boss's office than he caught me in the hall. I listened politely while he told me that he was going to have to pay overtime to the typists to make up for the time I cost him. He said a few other words which I'll not burden you with. When it was all off his chest, I asked him one question: "What would you have done in my place?" He grinned and said, "The same thing you did!"

The point I want to make is that the salesman, if doing his job, will be a disruptive force from time to time.

The next step in bringing this new person into your organization is to explain to all on the nuts-and-bolts side how the salesperson fits in. Your superintendents on the jobs will first take the attitude that the salesperson will be looking over their shoulder and reporting back to the boss. The best way to dispel this notion is for the salesperson to establish a rapport with the field people. The main reason he should be checking the jobs is to gather information to supply to the owner. And let the person running the job know this from both you and the salesperson. In the end it'll be up to the salesperson to integrate himself into the construction side of the company; and if he's good at his job, he'll do it with ease, since his job is dealing with people.

At the same time, if something is discovered that is going to affect the job and create problems with the owner, you'll be the first to know. If the superintendent gets his nose out of joint, then too bad—because your salesperson isn't going to stand by and let the owner get the wrong impression of him or the company. And if your salesman doesn't take this attitude, then you don't want him. Aggressive selling means being aggressive in more than one area.

There is a negative aspect to adding a sales department. Hopefully the salesperson will be able to make it disappear, but I personally think it never goes away altogether. Every sales force is resented to some degree by the rank and file. By management, no, because they recognize the contribution a sales force makes; but the people sitting behind the desk see a salesperson as a freewheeling good-time player taking a customer to lunch, then writing up an order with commissions amounting to what they make in a couple of months. Then these same people have to take care of the details to make sure the customer receives what he bought.

This resentment can really be present in a construction company with the people working on the site. It can begin with the person in charge and work right on down to the person with a hammer. Therefore, it's imperative to impress on your people in charge of the field just how important a job the salesperson has. Stress the concept of teamwork, that somebody has to build what the salespeople sell, and everyone is doing a vitally needed job. As I said, I don't think the resentment will ever go away completely. What you're working toward is making it an insignificant factor and obtaining money-making results from your employees.

It's going to be up to you, the employer, to put the salesperson in the picture. Afterward it'll be his job to make it work. The salesperson should report only to you or top management; no one else should have one thing to say about how he does his job. His position is comparable to a staff job where one has the authority to get his work done but has no one answering directly to him.

A SALESPERSON'S RESPONSIBILITY TO THE COMPANY

As much of an individual as the salesperson will probably be, and as much as he will dislike paperwork, there will be certain duties that you should expect. Proper records are a must. It's vital for you to know what's being worked on and what the near future holds for contracts. The salesperson should keep lead and prospect lists with all pertinent information for each listing, status of the prospect at the last meeting, and what is to be done before the next meeting. The time of the next meeting should be recorded also. This database will serve two purposes: it will give you, the employer, a handy printout to check to get a feel for what the salesperson is pursuing, and it will provide the salesperson with a printout that he can use to keep all the loose ends under control.

Along with the prospect data it's very important to spend some time every one to two weeks with your salesperson and go over the prospects as well as owner relations with jobs under construction. This exchange of information can be very helpful to both of you, and all concerned will stay up-to-date.

Don't try to tell the salesperson how he should go about selling. I've seen it happen. It's hard for employers not to express their ideas on how the salesman should be doing his job. Convey overall expectations, but not the details. That's what you're paying the salesperson to know. So fight the urge to do it.

Communications during the day are pretty much up to the salesperson. I've found it best that he check with the office at least three times and never less than twice. It will be a good idea to inform him that you want him always to take the time to call the office. Now let me say here that if your salesman is a pro, he'll check with the office, since the office will be a source of leads. Of course, a pocket beeper should prove very handy for all concerned.

While working, the construction salesperson must be on guard against his giving orders to the construction people. The pro knows how to handle the construction side of the business; but if you have someone coming on board who doesn't know very much about how a construction outfit operates, then tell him what not to do. I know how difficult it sometimes is to keep quiet. Nevertheless, the salesperson should go through the person in charge when he sees something out of line. Although the employees don't work for him, he is recognized as management; therefore, most likely the workers will accept orders from the salesperson. Thus, the salesperson will be undermining the superintendent, which is guaranteed to create friction between the sales and construction departments.

At some time everyone in the company will look on the salesperson as an errand person. It's only natural to say to let so-and-so drop it off. This can get completely out of hand. It is best for all concerned to make it clear that the salesperson is not a messenger. Now I'm the first to admit there will be occasional exceptions that are not inconvenient or are important enough to warrant it. What you must be on guard against is letting these exceptions become more frequent. You can tell the office people, and the salesman as well can say no to the office staff; but how about you? You are the one person the salesman works for, and he cannot refuse you; so it's mainly to you that I'm directing this advice. That salesman is worth too much to be used in this manner.

Do not expect the salesperson to keep regular office hours. He must be free to move where the business takes him on the time schedule that gets the job done. As I said before, he won't sell a thing in the office.

FINDING THE CONSTRUCTION SALESPERSON

You're wondering just how you locate a construction salesperson. You know the local employment agencies won't be able to help you. This is simply too specialized a field. So you're back to square one.

It's not really that difficult if you know where and what to look for. Check around with the competition and determine who's using salespeople. If there are some operating in your area, you can approach them to see if any will consider working for you. This is probably the fastest way to hire a salesperson. It can also be the most costly financially. Don't expect to bring in an experienced person for peanuts. The real problem is that there are not very many truly professional construction salespeople around, so chances are you'll have to create one.

If you list the people working for you, to think about who you can make a salesperson, you can be wasting your time. You might luck out and have someone on your payroll who can do the job, but it's a million-in-one shot. People with sales ability don't go to work for construction firms; nuts-and-bolts people work for contractors. Sales-minded people go where salespeople are used; so that's where you have to look.

The very best place to start your hunting expedition is within the construction industry itself as well as closely allied fields. I really don't recommend going outside this area because then the salespeople will know absolutely nothing of the construction industry.

You're looking for someone who's first and foremost a salesperson; that's the most important fact. Next you want this person to have a little knowledge of construction. He doesn't have to be an ex-superintendent, but it helps if he knows the difference between a concrete slab and a built-up roof. You can teach a good salesperson the necessary construction details.

This salesperson must be used to selling to the person in charge. An order taker who calls on purchasing agents is not for you. Your salesperson must understand why he has to deal with the person who signs the checks. Next it's very helpful if his background has been in larger-amount orders.

You want a person who's been in your area long enough to get established. So much of construction selling is who you know, and a newcomer will be at a decided disadvantage. A pro will pick up the details of the job quickly, but if you have to train a salesman to sell, then you're wasting your time.

Age is a factor that will affect sales. I have a stockbroker friend who was delighted when he started getting gray at the temples. He also tells the story about another broker who had a little gray added over his ears to look older. Whether the story is true or not, the point is well taken. There are several selling positions where maturity is a definite advantage, and construction sales is right at the top of the list. By far, the owners in the market for a new facility will be around 40 or older. These owners will be graduates of the business school of hard knocks. Confidence is a large slice of the selling pie, and these owners when talking about thousands of dollars prefer to talk with mature people.

The whole thing boils down to this: you're looking for a mature, established professional salesman who has experience selling construction related items in large-amount orders to the person who signs the check. And believe it or not, there are plenty of candidates running around in the construction industry and closely allied fields. Such as who, you ask? For openers, there are the factory reps for the building material manufacturers. Factory reps for the preengineered metal building manufacturers also fall into this category. Construction equipment salespeople also are possibilities. Any segment of the industry that uses salespeople can offer the good raw material for a construction salesperson.

In the allied fields there are factory reps for construction steel and other metal products. Industrial mill supply salespeople have an acceptable background. It's my opinion that the factory reps for building materials and metal building manufacturers are the most suitable source of good construction salespeople. They will have all the assets plus the added advantage of being used to running a sales territory on their own. To them, the home office is not across town but in the next state. To do a successful job, they have to be self-starters and able to work with very little supervision. Both of these qualities are absolutely essential to the results-getting construction salesperson. The only weakness with these reps is being established in your area. If you happen to find a person who's a hometown resident, then so much the better.

When you find one of these reps who does want to make a change, do your homework to find out why. You don't want to end up with some sales manager's problem child. You'll discover after talking with some of these people about changing jobs that the main reasons they are thinking about leaving is the desire to get off the road and wanting better opportunities to make more money. This will come home to them as they grow a little older, so their age is in your favor also.

The other areas I've mentioned are good salesman sources also, only not quite so good as the factor rep group; don't ignore these other sources. Matter of fact, one of the best construction salesmen I ever had the pleasure of knowing and working with came from a supplier of metal products. He was a prime example of why you have to hire a salesman first, with all else secondary. This guy was the best psychologist I've ever seen when it came to the prospect. He knew exactly what the prospect was going to say before he opened his mouth. He knew how to approach the job, and, better yet, he was a closer—he got the prospect's name on the dotted line. But he always had a little trouble with construction details. He never learned the correct names for all the details in some areas, such as electrical. For example, everything was a switch or a plug on the wall; never exactly what it was, simply a switch or a plug. He never could remember the difference between a column and a beam; and I'll bet right now if you asked him, he'd stop and think about it and then give you the wrong answer. After a while the two owners of the construction company that we both worked for gave up on him. If the prospects didn't mind, why should they? He just kept bringing in the contracts.

He was a fine salesman, always selling himself and his company, and on the lookout for leads twenty-four hours a day, seven days a week. This man was a super construction salesman who came from a metal products firm. He never mastered all the details on the construction side; and, what's more, he would tell you he didn't care. He was a salesman, not an engineer.

My point is that no matter where you find your salesman, he'd better be a salesman first.

SALES TERRITORIES WHERE TWO OR MORE SALESPEOPLE ARE INVOLVED

If you have or plan to have two or more salespeople, there should be definite areas of responsibility outlined. These established territories should be for leads and prospects, not actual construction sites. The territory is where the salesperson will work seeking leads and working on prospects. This includes incoming phone leads from the territory, as well. If the prospect is going to build and the site is in another person's territory, this situation will present no problem. The job belongs to the person who sold the job. This particular situation will come up because the prospect will be moving or in some cases adding a branch operation.

There will be conflicts from time to time involving leads that start in one area and end up in another territory. My personal feeling is that the lead in such a case should be passed on to the person who handles the territory. Or you can remain uninvolved and let the two salespeople work things out. You can't afford squabbling over leads. It's important that salespeople work together, for if you're doing any business at all, there'll be times when one has to cover for the other. Good relations in the construction sales department is a must.

You may feel the salespeople are mature and objective enough to work out the territories on their own. If so, let them, as long as the results are there. After all, results are what you're really interested in. My last bit of advice on this subject is not to create a very formal sales department with rigid guidelines. You want an efficient, fluid operation that does what has to be done to get the contract. When I was one of two salespeople working for a construction company, a lead came to my attention that was located in the other salesperson's territory. I got the lead because the party in question was a personal friend. I hesitated to pass the lead on to the other salesperson because I was the one who knew the party and the background, and he had come to me because we were friends. Therefore, I was the one with the best chance of having a contract signed. Also, it was a lead that was not out on the streets, and chances were the other salesperson would never have heard of the job. We talked it over. The first thing was to get the project for the firm; and since I had the inside track, he told me to sell and handle the job, which I did. Don't shackle your salespeople with rigid rules. Let them operate in an independent manner, as long as it gets the results.

SALESPERSON REMUNERATION

Remuneration in plain language is *money!* Nothing's free and 99% of the time you get exactly what you pay for. A good results-getting construction salesperson is not cheap. While hard numbers would be very difficult to present, I can nevertheless offer some general rules.

I advise against a straight salary. First, if the prospective salesperson wants a straight salary, mark him off your list. This is a good indicator of what kind of salesperson he is. A straight-salaried person is not aggressive enough for construction sales. It's too easy to become lazy when the money is always there. Of course, if he's not working, the lack of results will show it, and you can take proper action.

Most of the salesperson's money should come from commissions, and the salesperson, if he's the caliber you want, will prefer it this way. Rely on your own good common business sense about income. Just remember to maintain some control, don't put a limit on earned income, and never pay a straight salary.

Index

MIX
Papier aus verantwortungsvollen Quellen
Paper from responsible sources
FSC® C105338

If you have any concerns about our products,
you can contact us on
ProductSafety@springernature.com

In case Publisher is established outside the EU,
the EU authorized representative is:
Springer Nature Customer Service Center GmbH
Europaplatz 3, 69115 Heidelberg, Germany

Printed by Libri Plureos GmbH
in Hamburg, Germany